T0184899

Machine Learning in Medicine

Ton J. Cleophas • Aeilko H. Zwinderman

Machine Learning in Medicine

Part Two

by

TON J. CLEOPHAS, MD, PhD, Professor,
Past-President American College of Angiology,
Co-Chair Module Statistics Applied to Clinical Trials,
European Interuniversity College of Pharmaceutical Medicine, Lyon, France,
Department Medicine, Albert Schweitzer Hospital, Dordrecht, Netherlands,

AEILKO H. ZWINDERMAN, MathD, PhD, Professor,
President International Society of Biostatistics,
Co-Chair Module Statistics Applied to Clinical Trials,
European Interuniversity College of Pharmaceutical Medicine, Lyon, France,
Department Biostatistics and Epidemiology, Academic Medical Center,
Amsterdam, Netherlands

With the help from
HENNY I. CLEOPHAS-ALLERS, BChem

 Springer

Ton J. Cleophas
Sliedrecht
The Netherlands

Aeilko H. Zwinderman
Department of Epidemiology
    and Biostatistics
Academic Medical Center
Amsterdam
The Netherlands

ISBN 978-94-007-9512-9   ISBN 978-94-007-6886-4 (eBook)
DOI 10.1007/978-94-007-6886-4
Springer Dordrecht Heidelberg New York London

Preface

Machine learning is a novel discipline concerned with the analysis of large data and multiple variables. It involves computationally intensive methods, is currently mainly the domain of computer scientists, and is already commonly used in social sciences, marketing research, operational research, and applied sciences.

It is virtually unused in clinical research. This is probably due to the traditional belief of clinicians in clinical trials where multiple variables are equally balanced by the randomization process and are not further taken into account. In contrast, modern computer data files often involve hundreds of variables like genes and other laboratory values, and computationally intensive methods are required for their analysis.

The book *Machine Learning in Medicine* published in March 2013 by Springer, Dordrecht, the Netherlands, was written as an introduction to the medical community, and consists of 20 chapters relevant to clinical research. It addresses subjects like optimal scaling, neural networks, factor analysis, partial least squares, discriminant analysis, canonical analysis, and fuzzy modeling.

Why are these methods so convenient for clinical data analysis? This is so because (1) they can handle a large amount of or complex data is no problem, (2) they are more flexible than traditional statistical methods, (3) software programs based on these methods offer user friendly menus without a lot of syntax.

Prior to the completion of the first volume of this publication the authors came to realize that additional methods accounting the effects of potential biases of the reported methods were available. Also, some of the methods not reported so far might provide a better fit for some clinical data sets. These points were the main reasons for writing the current volume *Machine Learning in Medicine: Part Two*. Like the first volume it consists of 20 chapters, and it reviews methods like various clustering models, support vector machines, Bayesian networks, discrete wavelet analysis, genetic programming, association rule learning, and anomaly detection.

We as authors believe that, together with the first book, this book is unique in its field, because the medical community, although increasingly involved in large and complex data, is lacking adequate knowledge for in-depth analyses. Each chapter of the novel book is written in a way much similar to that of the first volume and

(1) will describe a novel method that has already been successfully applied in the authors' own research, (2) will provide the analysis of medical data examples, (3) will provide step-by-step analyses for the benefit of the readers, (4) will provide the commands to be given to the software programs applied (mostly SPSS), (5) can be studied without the need to consult other chapters, and (6) will be written in a explanatory way for a readership of mainly nonmathematicians.

We should add that the authors are well qualified in their field. Professor Zwinderman is president of the International Society of Biostatistics (2012–2015), and Professor Cleophas is past-president of the American College of Angiology (2000–2002). From their expertise they should be able to make adequate selections of modern methods for clinical data analysis for the benefit of physicians, students, and investigators. The authors have been working and publishing together for 14 years, and their research can be characterized as a continued effort to demonstrate that clinical data analysis is not mathematics but rather a discipline at the interface of biology and mathematics.

The authors are not aware of any textbook in the field of machine learning in medicine, and, therefore, believe that the current two-volume publication entitled *Machine Learning in Medicine Parts One* and *Two* does fill a need.

18 March 2013   Prof. Ton J. Cleophas
Lyon   Prof. Aeilko H. Zwinderman

Contents

Chapter 1
Introduction to Machine Learning Part Two

1 Summary

1.1 Machine Learning Is Convenient

Machine learning is convenient to clinical research, because (1) big and complex data is no problem, (2) it is more flexible than traditional statistical methods, (3) user friendly menus without a lot of syntax is, generally, offered by the various software programs.

1.2 Limitations

Limitations include: machine learning methods do not always meet scientific standards and one has to be cautious with predictions about future data from machine learning results.

1.3 Wonderful Methods

It may take a while before traditional null-hypothesis testing is replaced with machine learning methods in clinical research, but machine learning methods have already provided some wonderful methods relevant to clinical research.

T.J. Cleophas and A.H. Zwinderman, *Machine Learning in Medicine:*
Part Two, DOI 10.1007/978-94-007-6886-4_1,
© Springer Science+Business Media Dordrecht 2013

1.4 Important Methods Reviewed in Volume 1

Important methods reviewed in volume 1 are: optimal scaling, neural networks, factor analysis, partial least squares, discriminant analysis, canonical analysis, and fuzzy modeling.

1.5 Important Methods Reviewed in Volume 2

Important methods reviewed in volume 2 are: various clustering models, support vector machines, Bayesian networks, discrete wavelet analysis, genetic programming, association rule learning, anomaly detection.

2 Introduction, What Is so Convenient About Machine Learning

Machine learning is a discipline concerned with the statistical analysis of big data and data with many variables. The objectives of machine learning are just like those of traditional statistical analyses, namely (1) to make predictions, e.g., about the efficacy of novel treatments, and (2) to identify patterns, e.g., for the purpose of making health decisions. Machine learning methods is little used in clinical research. This is probably due to the traditional belief of clinicians in clinical trials where multiple variables are equally balanced by the randomization process and are not further taken into account. In contrast, modern computer data files often involve hundreds of variables like genes and other laboratory values, and computationally intensive methods are required for their analysis.

The methods of machine learning are much like those of traditional statistical methods, and include summary statistics, data plots, normal and distribution free models, and regression methods. Sometimes, novel methods in machine learning have their priors in traditional statistics, like entropy and threshold/shrinkage methods that have their priors in respectively logistic and Bayesian modeling.

What, then, is so convenient about machine learning? That is (I) that machine learning programs are better able than traditional methods to handle big and complex data without the problem that commands are not executed because of numerical problems and integer overflow. Also, (II) they are more flexible. For example, discrete wavelet learning (Chap. 19, vol (volume) 2), compared to traditional methods like Fourier regression and other methods for non linear data, provides five important advantages:

(1) You can determine which features in your data are important and which are not,
(2) It reveals aspects that other methods tend to miss,

(3) It always produces a better fit to the data than Fourier does,

(4) It can because of its better fit be more easily denoised,

(5) It can account constantly changing patterns.

Finally, (III) software, generally, consists of user friendly menus including validating and contrast tests and little need for a lot of syntax knowledge.

3 Limitations of Machine Learning, Advantages of Traditional Statistical Methods

The increased flexibility is somewhat offset by a reduced scientific rigor. Traditional statistical methods have been developed by statisticians with a mathematical background and aware of the limitations of their tools, while many machine learning workers, stemming from the fields of econo-, sociometry, and computer science, have been less so.

Traditional statistical methods are based on strict and consistent scientific rules, prior validations and goodness of fit tests. With machine learning this is also less so. For example, the presence of theoretical sampling distributions and other prior assumptions are rarely checked. The increased risk of type I and II errors due to multiple testing and unrandom sampling is not routinely taken into account. Chapter 20, vol 2 will explain the potential flaws of machine learning for validating diagnostic tests, and the need for common sense reasoning as a guide for machine learning workers.

Indeed, results from machine learning are not always valid, or, at least, do not always meet desired scientific standards. They should, generally, be considered exploratory rather than confirmatory. But they do, often, provide interesting novel concepts, and are, often, helpful in practice. Attempts have been made to improve scientific standards for machine learning, but this was not always successful. E.g., The Java Data Mining (JDM) Standards were launched in 2006, and withdrawn later the same year [1]. It is obvious, that one has to be cautious with the results of machine learning analyses for decisions about future data.

Some more advantages of traditional statistical analyses are given. First, for a better understanding of the assumed nature of the relationship between the x- and y-variables a parabolic or Fourier's or polynomial regression is helpful and sometimes better than machine learning methods that do not follow simple mathematical models, such as spline analysis, Loess analysis and wavelet analysis for non linear data (Chap. 11, vol 1). Second, as a matter of course, additional variance is included in more complex models. More complex models are ipso facto more at risk of power loss, particularly, with relatively small data. Third, as much with numerical procedures, more sophisticated methods accounting the effects of potential biases are, currently, generally, available. They may provide better fit for some data sets, but at the same time are at risk of power loss if for one reason or another the data do not fit well.

4   Machine Learning and Megadata

Machine learning is an important field in computer science. Computer science is also largely involved in the development of megacomputers, like the Bluegene computers and the Top 500 computers, that are able to store and process megabyte, like terabyte, data. Examples of megadata-systems are Facebook, eBay, Twitter etc. Such systems raise questions regarding privacy, legality, and ethics, but, at the same time, have already been able to readily provide clinicians with important medical (although uncensored) information, like Google has. Megadata-systems are, sometimes, named the next industrial revolution and the end of theory (or hypothesis) [2]. Megadata analyses are often analyzed using grid computing systems, rather than traditional statistical hypothesis tests.

Is this also relevant to clinical research? In clinical research the objective is generally to assess novel versus standard treatments, and a null-hypothesis test is used for confirmation, rather than a grid of data. The statistics for megadata and small data are not different. There are no special null-hypothesis tests for megadata. A problem is that traditional null-hypothesis tests of large data in clinical research are often rather unpleasant, not only because of computational difficulties, but also, because they are overpowered, meaning that you are apt to find (multiple) small effects that are statistically significant but clinically irrelevant. Traditional statistical tests are designed to find clinically relevant differences in relatively small samples.

And so, it may take a while before the traditional null-hypothesis test is dismissed in the field of clinical research, but, in the meantime, machine learning has already brought some wonderful methods that can be readily applied as a help to traditional clinical data analyses. The current 2 volumes review over 40 such methods, including examples to illustrate how they can be implemented in clinical research.

5   The Machine Learning Methods Already Reviewed in the First Volume of This Publication

1. Machine learning methods are still in a stage of development, and, therefore, hard to definitely classify. A classification as either supervised if an outcome variable is included (like with discriminant analysis, (Chap. 17, vol 1)) or unsupervised if not (like with hierarchical cluster analysis (Chap. 15, vol 1)), is useful, but does not cover the field by far. Other supervised machine learning methods involve multivariate techniques like artificial intelligence (multilayer perceptron modeling, Chap. 12, vol 1, radial basis function analysis, Chap. 13, vol 1), canonical regression (Chap. 18, vol 1), partial least squares (Chap. 16, vol 1), factor analysis (also factor analysis is a multivariate method because the factors are the outcomes) (Chap. 14, vol 1).
2. Some would prefer to favor machine learning for the assessment of special effects like interaction effect removal (partial correlations, Chap. 5, vol 1), adjusting

overdispersion (optimal scaling, Chaps. 3 and 4, vol 1), adjusting confounding (logistic regression for health profiling, Chap. 2, vol 1), adjusting time-dependent risk factors (multiple Cox regression with time-dependent predictors, Chap. 9, vol 1), seasonality identification (autocorrelations, Chap. 10, vol 1).

3. Others would do so for better sensitivity of testing, and would for the purpose apply mixed linear models (when within-subject differences are small) (Chap. 6, vol 1), non linear models (Chap. 11, vol 1), and principal components analysis (factor analysis, Chap. 14, vol 1).
4. Still others would do so for removing the flaws of other analyses like binary partitioning (Chap. 5, vol 1) removing the flaw of ROC (receiver operating characteristic) analyses that do not account the sample size of the positive and negative tests.
5. Machine learning can also be applied for avoiding the need for laborious validity/reliability assessments, and, e.g., item response modeling can be used for that purpose (Chap. 8, vol 1).
6. Machine learning can be applied for diagnostic testing with the use of a learning sample. Back propagation neural networks (Chap. 12, vol 1), radial basis neural networks (Chap. 13, vol 1), receiver operating characteristic modeling (Chap. 7, vol 1), and binary partitioning (Chap. 7, vol 1) can be used here.

6   The Basic Models for Machine Learning

Machine learning models may or may not be different from those of traditional statistics. Orthogonal linear modeling is the basis of both factor analysis and discriminant analysis. Distances, otherwise called proximities, are the basis of many cluster models, usual regression is the basis of methods like canonical, partial-regression, health-profiling, mixed linear modeling, time-dependent-Cox, optimal-scaling, autocorrelation, partial-least-squares, and most non linear models. Entropy of parent and internal nodes is the basis of binary partitioning. Connecting polynomes (second and third order functions) with first and second derivatives is the basis of spline analysis. Hidden layers filled out with data-based distribution-free results is the basis of artificial intelligence, and triangular fuzzy sets is the basis of fuzzy modeling.

7   The Machine Learning Methods to Be Reviewed in the Second Volume of This Publication

Already prior to the completion of the first volume the authors came to realize that additional methods accounting the effects of potential biases of the reported methods were available. Also, some of the methods not reported so far might provide better fit for some clinical data sets. These points were the main reasons for writing the book "Machine Learning in Medicine Part Two".

First, we will review methods that are more sensitive than traditional methods, e.g., multistage least square as compared to traditional least squares (Chap. 2, vol 2), logistic regression with the odds of disease as outcome as compared to concordance statistics (Chap. 6, vol 2), association rule learning as compared to simple regression (Chap. 11, vol 2), the odds ratio approach to quality of life for better precision than that of traditional summation of scores (Chap. 5, vol 2), support vector machines for modeling dependent variables with better sensitivity/specificity than traditional logistic regression (Chap. 16, vol 2).

Second, we will review methods that are extensions of traditional methods, like two-dimensional (Chap. 8, vol 2) and multidimensional clustering (Chap. 9, vol 2), support vector machines (Chap. 15, vol 2), anomaly detections (Chap. 10, vol 2), that can be considered as extensions of simple hierarchical clustering (Chap. 15, vol 1). Also correspondence analysis (Chap. 13, vol 2) is a 3×3 (or more) extension of simple 2×2 contingency table analysis. Autoregressive integrated moving average (ARIMA) (Chap. 14, vol 2) is an extension of simple autocorrelations (Chap. 10, vol 1), and so is discrete wavelet analysis of continuous wavelet analysis (Chap. 19, vol 2), multiple imputations of the more simple methods of dealing with missing data like regression imputation (Chap. 3, vol 2), Bayesian networks of the more basic regression models for causal relationships (Chap. 16, vol 2).

Third, we will review methods for which there is no alternative, like multidimensional scaling for assessing patients' personal meanings (Chap. 12, vol 2), validating surrogate endpoints (Chap. 7, vol 2), Bhattacharya modeling for finding hidden Gaussian subgroups in large populations (Chap. 4, vol 2), continuous sequential techniques (Chap. 18, vol 2), string/sequence mining of very large data (Chap. 17, vol 2).

We should add that, unlike traditional statistical methods, many machine learning methodologies consist of socalled unsupervised data, i.e., datasets without outcome variables. Instead of classification of the data according to their outcome values, the data are, then, classified according to their proximity, entropy, density, pattern. The Chaps. 4, 8, 9, 10 and 12 are examples.

8 Conclusions

Machine learning is convenient to clinical research, because big and complex data is no problem, machine learning is more flexible than traditional statistical methods, the software offers user friendly packages. Limitations include: machine learning methods do not always meet scientific standards and one has to be cautious with predictions about the results.

It may take a while before traditional null-hypothesis testing is dismissed in clinical research, but machine learning methods have provided already some wonderful methods relevant to clinical research.

Important machine learning methods reviewed in volume 1 of this publication are: optimal scaling, neural networks, factor analysis, partial least squares, discriminant analysis, canonical analysis, and fuzzy modeling. Important machine learning methods reviewed in the current volume 2 of this publication are: various clustering models, support vector machines, Bayesian networks, discrete wavelet analysis, genetic programming, association rule learning, anomaly detection.

References

1. Java data mining standards. www.amazon.com. 13 Feb 2013
2. Graham M (2012) Big data and the end of theory. The Guardian http://www.guardian.co.uk. 20 Feb 2013

Chapter 2
Two-Stage Least Squares

1 Summary

1.1 Background

Both path analysis and multistage least squares are adequate for simultaneously assessing both direct and indirect predictors. This makes interpretation less easy. Also, path analysis does not provide overall p-values.

1.2 Objective

To assess whether one method performs better than the other.

1.3 Methods

A real clinical data example of patients' non-compliance in a drug efficacy study was used. Data analysis was performed using SPSS statistical software.

1.4 Result

A two-path statistic of 0.46 is a lot better than the single-path statistic of 0.19 with an increase of 60.0 %. However, the result is expressed in a standardized way and without overall p-value, and so it is not easy to interpret. The p-value of two-stage least squares was (much) better than that of the simple linear regression p-value of non-compliance versus therapeutic efficacy (0.02 versus 0.10).

T.J. Cleophas and A.H. Zwinderman, *Machine Learning in Medicine:*
Part Two, DOI 10.1007/978-94-007-6886-4_2,
© Springer Science+Business Media Dordrecht 2013

1.5   Conclusions

1. Path analysis and multistage least squares are linear regression methods that are adequate for simultaneous assessment of direct and indirect effects of clinical predictors.
2. Multistage least squares is easier to interpret, because it provides unstandardized regression coefficients and an overall p-value.
3. Multistage least squares is at risk of overestimating the precision of the outcome.

2   Introduction

Multistage regression, otherwise called multistage least squares, was invented by Philip G. Wright, professor of economics at Harvard University, Cambridge MA, 1928 [1]. It uses instrumental variables for improved estimation of problematic (i.e., somewhat uncertain) predictors [2]. It is an alternative to traditional path analysis which assumes that predictor variables not only produce direct effects on the outcome but also indirect effects through affecting concomitant predictor variables. With path analysis usual regression coefficients can not be applied, because they have the same unit as the outcome variable. Instead, standardized regression coefficients have to be used. This makes interpretation less easy. Also, path analysis does not provide overall p-values. In the current chapter two stage least squares is explained as an alternative to traditional path analysis, using a real data example.

3   Example Path Analysis

Patients' non-compliance is a factor notoriously affecting the estimation of drug efficacy. An example is given of a simple evaluation study that assesses the effect of non-compliance (pills not used) on the outcome, efficacy of a novel laxative, with numbers of stools in a month as efficacy estimator (the y-variable). The data are in the first three columns of Table 2.1.

Table 2.2 gives the results of two linear regressions assessing the effects of counseling and non-compliance on therapeutic efficacy (upper table), and the effect of non-compliance on counseling (lower table). With $p = 0.10$ as cut-off p-value for statistical significance all of the effects are statistically significant. Non-compliance is a significant predictor of counseling, and at the same time a significant predictor of therapeutic efficacy. This would mean that non-compliance works two ways: it predicts therapeutic efficacy *directly* and *indirectly* through counseling. However, the indirect way is not taken into account in the usual single step linear regression. An adequate approach for assessing both ways simultaneously is path statistics.

Table 2.1 Example of a study of the effects of counseling and non-compliance on the efficacy of a novel laxative drug (*pt* patient no.)

Pt	Instrumental variable (z) Frequency counseling	Problematic predictor (x) Pills not used (non-compliance)	Outcome (y) Efficacy estimator of new laxative (stools/month)	Improved values of problematic predictor
1.	8	25	24	27.68
2.	13	30	30	30.98
3.	15	25	25	32.30
4.	14	31	35	29.00
5.	9	36	39	28.34
6.	10	33	30	29.00
7.	8	22	27	27.68
8.	5	18	14	25.70
9.	13	14	39	30.98
10.	15	30	42	32.30
11.	11	36	41	29.66
12.	11	30	38	29.66
13.	12	27	39	30.32
14.	10	38	37	29.00
15.	15	40	47	32.30
16.	13	31	30	30.98
17.	12	25	36	30.32
18.	4	24	12	25.04
19.	10	27	26	29.00
20.	8	20	20	27.68
21.	16	35	43	32.96
22.	15	29	31	32.30
23.	14	32	40	31.64
24.	7	30	31	27.02
25.	12	40	36	30.32
26.	6	31	21	26.36
27.	19	41	44	34.94
28.	5	26	11	25.70
29.	8	24	27	27.68
30.	9	30	24	28.34
31.	15	20	40	32.30
32.	7	31	32	27.02
33.	6	29	10	26.36
34.	14	43	37	31.64
35.	7	30	19	27.02

Essentially, path analysis assumes two effects, (1) the effect of non-compliance on efficacy, and (2) the effect of non-compliance through counselling on efficacy. These two effects can be, simply, added up, and can be used to cover the joint effects of non-compliance on the efficacy. If we want to add up the effects of both variables,

Table 2.2 The results of two linear regressions assessing (upper table) the effects of counseling and non-compliance on therapeutic efficacy, and (lower table) the effect of non-compliance on counseling

Coefficients[a]

Model	Unstandardized coefficients		Standardized coefficients	t	Sig.
	B	Std. error	Beta		
1 (Constant)	2.270	4.823		.471	.641
Counselling	1.876	.290	.721	6,469	.000
Non-compliance	.285	.167	.190	1,705	.098

Coefficients[b]

Model	Unstandardized coefficients		Standardized coefficients	t	Sig.
	B	Std. Error	Beta		
1 (Constant)	4.228	2.800		1.510	.141
Non-compliance	.220	.093	.382	2.373	.024

[a]Dependent variable: ther eff
[b]Dependent variable: counseling

their units will have to be the same. Standardizing the data is the solution. For that purpose both the values and their variance are divided by their own variance.

$$\text{Values} \quad \text{(variance of values)}$$

$$\frac{\text{Values}}{\text{variance}} \quad \frac{\text{(variance of values)}}{\text{variance}}$$

The "values/variance" terms are called the standardized values. They have a variance of 1, which is very convenient for the calculations. In the analysis all regressions are performed on standardized values, giving rise to standardized regression coefficients, which can be simply added up, since they now have the same unit, while variances (equaling 1) no longer have to be taken into account. The standardized regression coefficients are calculated by many software programs. SPSS statistical software [3] routinely reports the standardized regression coefficients together with the usual regression coefficients (Table 2.2).

With simple linear regression the standardized regression coefficient is equal to the r-value. A path diagram must be constructed with arrows for indicating supposed causal effect paths. Figure 2.1 summarizes the supposed effects for our example: (a) non-compliance causes an effect on efficacy, and (b) non-compliance causes an additional effect on efficacy though counseling. The standardized regression coefficients are added to the arrows (Fig. 2.1). Then, they are added up according to:

$$0.19 + 0.38 \times 0.72 = 0.46$$

This result is expressed as the path statistic, and equals the sum of the standardized regression coefficients. Its magnitude is sometimes interpreted similarly to the overall r-square value of a multiple regression, but this is not entirely correct.

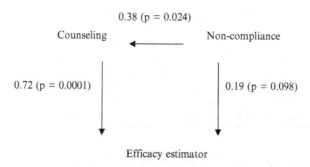

Fig. 2.1 Path diagram of study assessing the direct and indirect effects of non-compliance on efficacy of a new laxative

Unlike r-square values, standardized regression coefficients and their sums can be somewhat larger than 1.0. Also negative indirect factors are sometimes produced, reducing the magnitude of the add-up sums. Yet, the interpretation is pretty much the same. The larger the result of your path statistic, the better the independent variable predicts the dependent one. And, so, the two-path statistic of 0.46 is a lot better than the single-path statistic of 0.19 with an increase of 60.0 %.

4 Multistage Least Squares Method

Instead of path analysis, multistage least square is another possibility for the analysis of the above study. Also this method requires both a significant correlation between the predictors and the outcome in the multiple regression, and a significant correlation between the predictors. The simplest version, called 2 stage least squares (2SLS), is available in the regression module of SPSS [3], and is adequate for most data files. Its theoretical basis is slightly different from that of path analysis. Path analysis assumes causal pathways. The 2SLS method assumes that the independent variable (x-variable), here traditionally called the exogenous variable, is problematic. Problematic means that it is somewhat uncertain. If an additional variable can be argued to provide additional information about a problematic variable, then it may be worthwhile to include it in the analysis.

The variable counseling in the above example may, indeed, cause improvement of patients' compliance and, thus, indirectly improve the outcome variable. In the 2SLS model counseling will be used as an instrumental variable, for the purpose of reducing the uncertainty of the problematic variable non-compliance according to the underneath model:

y = outcome variable (drug efficacy)
x = problematic variable (non-compliance)
z = instrumental variable (counseling)

Table 2.3 Results of the 2LS analysis of the data from Table 2.1 (predictor = explanatory = non-compliance)

Model Description

		Type of Variable
Equation 1	VAR00001	dependent
	VAR00003	predictor
	VAR00002	instrumental

MOD_1

Coefficients

		Unstandardized Coefficients		Beta	t	Sig.
		B	Std. Error			
Equation 1	(Constant)	-61,095	37,210		-1,642	,110
	VAR00003	3,113	1,256	2,078	2,478	,019

1st stage

x = intercept + regression coefficient times z

With the help of the calculated intercept and regression coefficient from the above simple linear regression analysis improved x-values are calculated e.g. for patient 1

$x_{improved}$ = intercept + regression coefficient times 8 = 27.68

The right column of the Table 2.1 gives the improved x-values.
2nd stage

y = intercept + regression coefficient times improved x-values.

SPSS is used for analysis [3].

Command: Analyze....Regression....2 Stage Least Squares....Dependent: therapeutic efficacy....Explanatory: non-compliance.... Instrumental: counseling ....OK.

The Table 2.3 shows the results of the 2LS method. As expected the final p-value is (much) smaller (0.019 versus 0.098) than the simple linear regression p-value of the effect of non-compliance on therapeutic efficacy.

5   Discussion

The present chapter uses data of an efficacy study of a new treatment to explain path analysis and multistage least squares. Both are multistep statistical methods that are adequate for simultaneous assessment of direct and indirect effects. They are very successful in economics, but rarely used in clinical research. The current chapter shows that they can be readily applied to clinical efficacy studies, and enable to make better predictions from the data than does usual linear regression.

Some limitations of multistep regression have to be mentioned. First, instrumental variables may be weak predictors. In order to exclude weak predictors multistage regression should always be preceded by the usual multiple regression: only relatively strong predictor variables significantly predicting the outcome variables can be included. Second, also significant correlations between the predictor variables are required. At the same time, however, they must not be too strong. An r-value > 0.85 indicates the presence of collinearity, which is an important validity criterion of multiple regression.

We should add, that multistage least squares has the advantage compared to path analysis that it uses unstandardized b-values and provides an overall p-value. However, it assumes that instrumental variables are uncorrelated with the error terms of the problematic variable. If this is not warranted, precision of the outcome statistics is likely to be somewhat overestimated.

6 Conclusions

1. Path analysis and multistage least squares are linear regression methods that are adequate for simultaneous assessment of direct and indirect effects of clinical predictors.
2. Multistage least squares is easier to interpret, because it provides unstandardized regression coefficients and an overall p-value.
3. Multistage least squares is at risk of overestimating the precision of the outcome.

References

1. Wright PG (1928) The tariff on animal and vegetable oils. MacMillan, New York, Ed by Wright PC
2. Angrist JD, Krueger AB (2001) Instrumental variables and the search for identification: from supply and demand to natural experiments. J Econ Perspect 15:69–85
3. SPSS statistical software. www.spps.com. 25 Dec 2012

Chapter 3
Multiple Imputations

1 Summary

1.1 Background

In clinical research missing data are common. Imputated data are not real data, but constructed values that should increase the sensitivity of testing. Regression imputation may be more sensitive than means and hot deck imputation. Multiple imputations may be scientifically the best option, because uncertainty of the missing data is fully taken into account.

1.2 Objective and Methods

To assess the performance of all imputation methods and test them against one another in a single 35 patient dataset with five values randomly removed.

1.3 Results

The test statistics after regression-substitution were much better than those after the other two methods with F-values of 44.1 versus (vs) 29.4 and 31.0, and t-values of 7.6 vs 5.6 and 5.7, and 3.0 vs 1.7 and 1.8, but overstated the sensitivity of testing, because it produced even better statistics than the full dataset (without missing data) did. In contrast, the multiple imputations method provided test statistics that were virtually identical to those of the full dataset.

T.J. Cleophas and A.H. Zwinderman, *Machine Learning in Medicine:*   17
Part Two, DOI 10.1007/978-94-007-6886-4_3,
© Springer Science+Business Media Dordrecht 2013

1.4   Conclusions

The method of regression imputation was very sensitive, but overstated the sensitivity of testing, while the method of multiple imputations produced results virtually identical to those of the dataset without missing values.

2   Introduction

The imputation of missing data using mean values or values of the "closest neighbor observed" (sometimes called hot deck imputation), has been routinely carried out on demographic data files since 1960 [1]. The appointment of congressional seats and other political decisions have been partly based on it [2], and president Obama is having the White House use it again in its 2010 census [1]. Also in clinical research missing data are common, but compared to demographics, clinical research produces generally smaller files, making a few missing data more of a problem than it is with demographic files. As an example, a 35 patient data file of three variables consists of $3 \times 35 = 105$ values if the data are complete. With only five values missing (one value missing per patient) five patients will not have complete data, and are rather useless for the analysis. This is not 5 % but 15 % of this small study population of 35 patients. An analysis of the remaining 85 % patients is likely not to be powerful to demonstrate the effects we wished to assess. This illustrates the necessity of data imputation. Apart from the above two methods for data imputation regression-substitution is employed in clinical research. In principle, the blanks are replaced with the best predicted values from a multiple linear regression-equation obtained from the data available.

Imputated values are, of course, not real data, but constructed values that should increase the sensitivity of testing the data by increasing their fit to some analytical model chosen. Sensitivity is often expressed as the magnitude of the test statistic "mean value/SE (its standard error)". If a few values are imputed in a sample, then the SE will generally decrease, while the mean value might increase or decrease depending on the fit of the constructed values. Regression-substitution [3] may be more sensitive than the other two methods, because the relationship with all of the other values from the file is more closely taken into account. On the other hand, sensitivity may be overstated, because the best fit missing values are used without accounting their uncertainty. Uncertainty of the missing data is probably best taken into account by still another method, called multiple imputations: missing values are replaced with simulated versions using Monte Carlo modeling and their pooled results are in the final analysis [4, 5]. In the current chapter we will use the above 35 patient example to compare all of the four methods. Step-by-step analyses will be given.

3 Current Methods for Missing Data Imputation

Three traditional methods of data imputation commonly used are compared. The first one, a very old procedure, is to substitute the missing data of a variable with the mean value from that same variable. The problem with this approach is that no new information is given (the overall mean will remain unchanged after the imputations), but the standard error is reduced, and, thus, precision is overstated. The second method is often called the closest neighbor datum substitution. This method is, otherwise, called hot deck imputation, a term dating back to the storage of data on punched cards. The closest neighbor observation is found by subtracting the data of the patients with the missing data from those without the missing data one by one. The add-up sums of the smallest differences will unmask the closest neighbor. The problem with this approach is that both the patient and his/her closest neighbor may be outliers, and not provide the best fit for the data. As a third method, regression-substitution is possible. The incomplete data are used to calculate the best fit equation. For example, the best fit equation may look like:

$$y = 14 + 1.8\,x$$

If in a particular case the x-value is missing, then we can use the y-value to find the best fit x value. With $y = 42$, x should equal 16. Then we can impute the value 16 at the place of the missing datum. This method has the advantage, compared to the mean method, that the imputated datum is in some way connected with information from all of the other data. However, this conclusion is only true if the regression coefficient, the b-value, is statistically significant. Sensitivity of regression imputation may be overstated, because the best fit missing values are imputed without accounting their uncertainty (here expressed as standard error). Uncertainty of the missing data is probably best taken into account by multiple imputations: missing values are replaced with simulated versions using Monte Carlo modeling with their pooled results in the final analysis. A step-by-step analysis will be given.

4 Example

Thirty-five patients with constitutional constipation are treated in a crossover study with a standard laxative bisacodyl and a new compound using the numbers of stool within 1 month time as the main outcome variable. We wish to determine whether the efficacy of the standard laxative is a significant predictor of the efficacy of the new compound, and, also, whether age is a significant concomitant predictor. SPSS 17.0 [6] was used for the linear regression analysis. Table 3.1 shows the data file.

Table 3.1 Complete data file of 35 patients, the first and second variable indicate respectively numbers of stool on a new and a standard laxative (bisacodyl), the third variable indicates the patients' ages (*lax* laxative)

New lax	Bisacodyl	Age
24.00	8.00	25.00
30.00	13.00	30.00
25.00	15.00	25.00
35.00	10.00	31.00
39.00	9.00	36.00
30.00	10.00	33.00
27.00	8.00	22.00
14.00	5.00	18.00
39.00	13.00	14.00
42.00	15.00	30.00
41.00	11.00	36.00
38.00	11.00	30.00
39.00	12.00	27.00
37.00	10.00	38.00
47.00	18.00	40.00
30.00	13.00	31.00
36.00	12.00	25.00
12.00	4.00	24.00
26.00	10.00	27.00
20.00	8.00	20.00
43.00	16.00	35.00
31.00	15.00	29.00
40.00	14.00	32.00
31.00	7.00	30.00
36.00	12.00	40.00
21.00	6.00	31.00
44.00	19.00	41.00
11.00	5.00	26.00
27.00	8.00	24.00
24.00	9.00	30.00
40.00	15.00	20.00
32.00	7.00	31.00
10.00	6.00	23.00
37.00	14.00	43.00
19.00	7.00	30.00

Table 3.2 gives the data file after randomly removing five values from the file. Table 3.3 gives the statistics of the full data, the missing data file, and the four imputation models. Both the standard laxative and the age were significant predictors with p-values of 0.0001 and 0.048. The removal of five values reduced the sensitivity of testing. The t-value of bisacodyl (B_1) fell from 6.3 to 5.9 (p-value 0.0001–0.0001), of age (B_2) from 2.0 to 1.7 (p-value from 0.048 to 0.101).

Table 3.2 The data file from
Table 3.1 with five values
randomly removed (*lax*
laxative)

New lax	Bisacodyl	Age
24.00	8.00	25.00
30.00	13.00	30.00
25.00	15.00	25.00
35.00	10.00	31.00
39.00	9.00	
30.00	10.00	33.00
27.00	8.00	22.00
14.00	5.00	18.00
39.00	13.00	14.00
42.00		30.00
41.00	11.00	36.00
38.00	11.00	30.00
39.00	12.00	27.00
37.00	10.00	38.00
47.00	18.00	40.00
	13.00	31.00
36.00	12.00	25.00
12.00	4.00	24.00
26.00	10.00	27.00
20.00	8.00	20.00
43.00	16.00	35.00
31.00	15.00	29.00
40.00	14.00	32.00
31.00		30.00
36.00	12.00	40.00
21.00	6.00	31.00
44.00	19.00	41.00
11.00	5.00	26.00
27.00	8.00	24.00
24.00	9.00	30.00
40.00	15.00	
32.00	7.00	31.00
10.00	6.00	23.00
37.00	14.00	43.00
19.00	7.00	30.00

The t-values are estimators of the strength of association of the separate x-variables with the y-variable, and their magnitudes were much larger for the regression-substitution than for means and hot deck imputation, 7.6 vs 5.6 and 5.7, and 3.0 vs 1.7 and 1.8.

Just like mean imputation and hot deck imputation, regression-substitution considerably changed both the B and the SE values of the data's regression equation. In the given example it can be observed, however, that the t-values of

Table 3.3 The regression coefficients and their p-values obtained using different methods of data imputation

B_1 bisacodyl	SE_1	t	Sig	B_2 age	SE_2	t	Sig
Full data							
1.82	0.29	6.3	0.0001	0.34	0.16	2.0	0.048
5 % Missing data							
1.89	0.32	5.9	0.0001	0.31	0.19	1.7	0.101
Means imputation							
1.82	0.33	5.6	0.0001	0.33	0.19	1.7	0.094
Hot deck imputation							
1.77	0.31	5.7	0.0001	0.34	0.18	1.8	0.074
Regression imputation							
1.89	0.25	7.6	0.0001	0.31	0.10	3.0	0.005
Multiple imputations							
1.84	0.31	5.9	0.0001	0.32	0.19	1.7	0.097

B regression coefficient, *SE* standard error, *t* t-value, *Sig* p-value

regression-substitution equation were larger than those of the full data's equation, 7.6 and 3.0 vs 6.3 and 2.0. The sensitivity of testing was larger than that of the full data, and it was, thus, overstated. Artificially changing the error as implemented by the add-on module Missing Value Analysis of SPSS [7] may be needed in this situation. In order to perform the multiple imputation method the SPSS add-on module "Missing Values" was used. First, the pattern of the missing data must be checked using the command "analyze pattern". If the missing data are equally distributed and no "islands" of missing data exist, the model will be appropriate.

The following commands are needed:

Transform.... random number generators.... Analyze.... multiple imputations.... impute missing data....(the imputated data file must be given a new name e.g. "study name imputated").

Five or more times a file is produced by the software program in which the missing values are replaced with simulated versions using Monte Carlo methods (Table 3.4). In our example the variables are continuous and, thus, need no transformation. If you run a usual linear regression of the summary of your "imputated" data files, then the software will automatically produce pooled regression coefficients instead of the usual regression coefficients. The multiple imputation method produced a much larger p-value for the predictor age than the regression imputation did, and the result was, thus, less overstated than it was with regression imputation. Actually, the result was similar to that of mean and hot deck imputation, and statistical significance at $p < 0.05$ was not obtained (Table 3.3). Why then do it anyway. The argument is that, with the multiple imputation method, the imputed values are not used as constructed real values, but rather as a device for representing missing data uncertainty. Also, in the given example, unlike regression imputation, it did not seem to overstate the sensitivity of testing (Table 3.3, p-values regression imputation versus multiple imputation 0.005 versus 0.097).

5 Discussion

The current chapter suggests that for limited and incidentally occurring missing data, multiple regression-substitution, that uses regression equations with statistically significant predictors, is a more sensitive method of data imputation than the traditional means and hot deck imputation are.

The International Conference of Harmonisation Guidance on General Considerations for Clinical Trials recognized a special case of missing data being the loss of patients due to progression of disease, death, or cumulative drug toxicity. The socalled LOCF [8] method (last observation carried forward) was recommended in this situation. An intention to treat analysis can, then, be performed on all of the data with the argument that the last observation may be the best possible prediction of what the observation would have been, had the patient been followed. This solution may be appropriate for its purpose, but not for incidentally occurring missing data during a trial due, e.g., to equipment dysfunction or the patients' inclination not to report at some moment.

Instead of the three methods discussed in this paper, two more modern methods for data imputation are possible. The first one is the maximum likelihood approach using log likelihood ratio tests, that are based, just like linear regression, on normal distributions, and are a bit more sensitive than conventional t- or ANOVA tests as applied in linear regression. SPSS statistical software has it in its add-on module Missing Value Analysis [7]. SPSS adds here a bit of random error to each substitution, a rather arbitrary procedure. The second one is multiple imputations. For that purpose Little and Rubin applied multiple hot deck imputations instead of a single one, and used the pooled result of them for final data analysis [9]. This method may provide better sensitivity than single hot deck imputation. Monte Carlo Markow simulation models are used for data generation (see Chap. 57 for the description of the principles as applied) [10]. The multiple imputation models are rather complex, and, again, bits of random error are added. Also, the result is less spectacular than that of regression imputation. Why then do it anyway? The argument is that, with the multiple imputation method, the imputed values are not used as constructed real values, but rather as a device for representing missing data uncertainty. This approach is a safe and probably, scientifically, better alternative to the standard methods. In the given example, unlike regression imputation, it did not seem to overstate the sensitivity of testing (Table 3.3, p-values regression imputation versus multiple imputation 0.005 versus 0.097).

6 Conclusions

In clinical research missing data are common. Imputed data are not real data, but constructed values that should increase the sensitivity of testing. Regression imputation may be more sensitive than means and hot deck imputation. Multiple

Table 3.4 Missing data file and five imputated data files (35 patients) produced by the SPSS add-

File 1			File 2			File 3		
New lax	Bisacodyl	Age	New lax	Bisacodyl	Age	New lax	Bisacodyl	Age
24.00	8.00	25.00	24.00	8.00	25.00	24.00	8.00	25.00
30.00	13.00	30.00	30.00	13.00	30.00	30.00	13.00	30.00
25.00	15.00	25.00	25.00	15.00	25.00	25.00	15.00	25.00
35.00	10.00	31.00	35.00	10.00	31.00	35.00	10.00	31.00
39.00	9.00		39.00	9.00	33.77	39.00	9.00	21.49
30.00	10.00	33.00	30.00	10.00	33.00	30.00	10.00	33.00
27.00	8.00	22.00	27.00	8.00	22.00	27.00	8.00	22.00
14.00	5.00	18.00	14.00	5.00	18.00	14.00	5.00	18.00
39.00	13.00	14.00	39.00	13.00	14.00	39.00	13.00	14.00
42.00		30.00	42.00	15.10	30.00	42.00	15.99	30.00
41.00	11.00	36.00	41.00	11.00	36.00	41.00	11.00	36.00
38.00	11.00	30.00	38.00	11.00	30.00	38.00	11.00	30.00
39.00	12.00	27.00	39.00	12.00	27.00	39.00	12.00	27.00
37.00	10.00	38.00	37.00	10.00	38.00	37.00	10.00	38.00
47.00	18.00	40.00	47.00	18.00	40.00	47.00	18.00	40.00
	13.00	31.00	23.86	13.00	31.00	39.02	13.00	31.00
36.00	12.00	25.00	36.00	12.00	25.00	36.00	12.00	25.00
12.00	4.00	24.00	12.00	4.00	24.00	12.00	4.00	24.00
26.00	10.00	27.00	26.00	10.00	27.00	26.00	10.00	27.00
20.00	8.00	20.00	20.00	8.00	20.00	20.00	8.00	20.00
43.00	16.00	35.00	43.00	16.00	35.00	43.00	16.00	35.00
31.00	15.00	29.00	31.00	15.00	29.00	31.00	15.00	29.00
40.00	14.00	32.00	40.00	14.00	32.00	40.00	14.00	32.00
31.00		30.00	31.00	13.60	30.00	31.00	10.37	30.00
36.00	12.00	40.00	36.00	12.00	40.00	36.00	12.00	40.00
21.00	6.00	31.00	21.00	6.00	31.00	21.00	6.00	31.00
44.00	19.00	41.00	44.00	19.00	41.00	44.00	19.00	41.00
11.00	5.00	26.00	11.00	5.00	26.00	11.00	5.00	26.00
27.00	8.00	24.00	27.00	8.00	24.00	27.00	8.00	24.00
24.00	9.00	30.00	24.00	9.00	30.00	24.00	9.00	30.00
40.00	15.00		40.00	15.00	27.38	40.00	15.00	33.62
32.00	7.00	31.00	32.00	7.00	31.00	32.00	7.00	31.00
10.00	6.00	23.00	10.00	6.00	23.00	10.00	6.00	23.00
37.00	14.00	43.00	37.00	14.00	43.00	37.00	14.00	43.00
19.00	7.00	30.00	19.00	7.00	30.00	19.00	7.00	30.00

imputations may be scientifically the best option, because uncertainty of the missing data is fully taken into account.

This chapter was to assess the performance of all imputation methods and test them against one another in single 35 patient dataset with five values randomly removed.

on module "Missing Vales" using the command "multiple imputations"

File 4			File 5			File 6		
New lax	Bisacodyl	Age	New lax	Bisacodyl	Age	New lax	Bisacodyl	Age
24.00	8.00	25.00	24.00	8.00	25.00	24.00	8.00	25.00
30.00	13.00	30.00	30.00	13.00	30.00	30.00	13.00	30.00
25.00	15.00	25.00	25.00	15.00	25.00	25.00	15.00	25.00
35.00	10.00	31.00	35.00	10.00	31.00	35.00	10.00	31.00
39.00	9.00	31.22	39.00	9.00	27.82	39.00	9.00	34.89
30.00	10.00	33.00	30.00	10.00	33.00	30.00	10.00	33.00
27.00	8.00	22.00	27.00	8.00	22.00	27.00	8.00	22.00
14.00	5.00	18.00	14.00	5.00	18.00	14.00	5.00	18.00
39.00	13.00	14.00	39.00	13.00	14.00	39.00	13.00	14.00
42.00	15.07	30.00	42.00	17.00	30.00	42.00	18.04	30.00
41.00	11.00	36.00	41.00	11.00	36.00	41.00	11.00	36.00
38.00	11.00	30.00	38.00	11.00	30.00	38.00	11.00	30.00
39.00	12.00	27.00	39.00	12.00	27.00	39.00	12.00	27.00
37.00	10.00	38.00	37.00	10.00	38.00	37.00	10.00	38.00
47.00	18.00	40.00	47.00	18.00	40.00	47.00	18.00	40.00
31.51	13.00	31.00	34.53	13.00	31.00	48.42	13.00	31.00
36.00	12.00	25.00	36.00	12.00	25.00	36.00	12.00	25.00
12.00	4.00	24.00	12.00	4.00	24.00	12.00	4.00	24.00
26.00	10.00	27.00	26.00	10.00	27.00	26.00	10.00	27.00
20.00	8.00	20.00	20.00	8.00	20.00	20.00	8.00	20.00
43.00	16.00	35.00	43.00	16.00	35.00	43.00	16.00	35.00
31.00	15.00	29.00	31.00	15.00	29.00	31.00	15.00	29.00
40.00	14.00	32.00	40.00	14.00	32.00	40.00	14.00	32.00
31.00	8.20	30.00	31.00	12.92	30.00	31.00	11.83	30.00
36.00	12.00	40.00	36.00	12.00	40.00	36.00	12.00	40.00
21.00	6.00	31.00	21.00	6.00	31.00	21.00	6.00	31.00
44.00	19.00	41.00	44.00	19.00	41.00	44.00	19.00	41.00
11.00	5.00	26.00	11.00	5.00	26.00	11.00	5.00	26.00
27.00	8.00	24.00	27.00	8.00	24.00	27.00	8.00	24.00
24.00	9.00	30.00	24.00	9.00	30.00	24.00	9.00	30.00
40.00	15.00	33.20	40.00	15.00	31.69	40.00	15.00	34.93
32.00	7.00	31.00	32.00	7.00	31.00	32.00	7.00	31.00
10.00	6.00	23.00	10.00	6.00	23.00	10.00	6.00	23.00
37.00	14.00	43.00	37.00	14.00	43.00	37.00	14.00	43.00
19.00	7.00	30.00	19.00	7.00	30.00	19.00	7.00	30.00

The test statistics after regression imputation were much better than those after the other two methods with F-values of 44.1 versus (vs) 29.4 and 31.0, and t-values of 7.6 vs 5.6 and 5.7, and 3.0 vs 1.7 and 1.8, but overstated the sensitivity of testing because it produced even better statistics than the full dataset without missing data did. In contrast, the method of multiple imputations provided test statistics that were virtually identical to those of the full dataset.

References

1. Anonymous. Hot deck imputation. http://www.conservapedia.com/Hot-Deck_Imputation. Accessed 30 Aug 2010
2. Anonymous (2001) Utah v Evans, 182 F. supp. 2d 1165
3. Haitovsky Y (1968) Missing data in regression analysis. J R Stat Soc 3:2–3
4. Feingold M (1982) Missing data in linear models with correlated errors. Commun Stat 11:2831–2833
5. Kshirsagar AM, Deo S (1989) Distribution of the biased hypothesis sum of squares in linear models with missing observations. Commun Stat 18:2747–2754
6. IBM SPSS statistics 17.0. www.spss.com/software/statistics/chnages.htm. Accessed 03 Sept 2010
7. IBM SPSS missing values. www.spss.com/software/statistics/missing-values/. Accessed 03 Sept 2010
8. Chi GY, Jin K, Chen G (2003) Some statistical issues of relevance to confirmatory trials; Statistical bias. In: Lu Y, Fang J (eds) Advanced medical statistics. World Scientific, River Edge, pp 523–579
9. Little RJ, Rubin DB (1987) Statistical analysis with missing data. Wiley, New York
10. Scheuren F (2005) Multiple imputation: how it began and continues. Am Stat 59:315–319

Chapter 4
Bhattacharya Analysis

1 Summary

1.1 Background

Bhattacharya modeling is a Gaussian method recommended by the Food and Agricultural Organization of the United Nations Guidelines for searching the ecosystem for hidden Gaussian subgroups in large populations. It is rarely used in clinical research.

1.2 Objective

To investigate the performance of Bhattacharya modeling for clinical data analysis.

1.3 Methods

Using as examples simulated vascular lab scores we assessed the performance of the Bhattacharya method. SPSS statistical software is used.

1.4 Results

1. The Bhattacharya method better fitted the data from a single sample than did the usual Gaussian curve derived from the mean and standard deviation with 15 versus 9 cuts.

T.J. Cleophas and A.H. Zwinderman, *Machine Learning in Medicine: Part Two*, DOI 10.1007/978-94-007-6886-4_4, © Springer Science+Business Media Dordrecht 2013

2. Bhattacharya models demonstrated a significant difference at $p < 0.0001$ between the data from two parallel-groups, while the usual t-test and Mann–Whitney test were insignificant at $p = 0.051$ and 0.085.
3. Bhattacharya modeling of a histogram suggestive of certain subsets identified three Gaussian curves.

1.5   Conclusions

Bhattacharya analysis can be recommended in clinical research for the purpose of (1) unmasking normal values of diagnostic tests, (2) improving the p-values of data testing, and (3) objectively searching Gaussian subsets in the data.

2   Introduction

In 1967 Bhattacharya, a biologist from India, presented a method for identifying juvenile-fish subgroups from random samples [1]. By now this test, based on Gaussian curves, has become a key-method for the analysis and sustainability of this important resource in the eco-system, and is recommended by the Food and Agricultural Organization of the United Nations Guidelines [2]. As Gaussian curves are the mainstream not only with fish population research, but also with clinical data, it is peculiar, that, so far, this method has not been widely applied in clinical research. When searching Pub Med we only found a few clinical-laboratory studies [3–6], epidemiological [7, 8] and genetic studies [9, 10]. Bhattacharya modeling unmasks Gaussian curves, as present in the data, and removes outlier frequencies. In clinical research it could be used

1. For unmasking normal values of diagnostic tests,
2. For improving the p-values of data testing, and
3. For objectively searching subsets in your data.

The current chapter uses as examples simulated vascular lab scores to investigate the performance of Bhattacharya modeling as compared to standards methods, and was written to acquaint the clinical research community with this novel method.

3   Unmasking Normal Values

Vascular laboratories often define their estimators of peripheral vascular disease according to add-up scores of ankle, thigh, calf, and toe pressures. Figure 4.1 upper graph and Table 4.1 left two columns give as an example the frequency distribution of such scores in La Fontaine stage I patients. Normal values, otherwise called reference values, customarily present the central 95 % of the values obtained

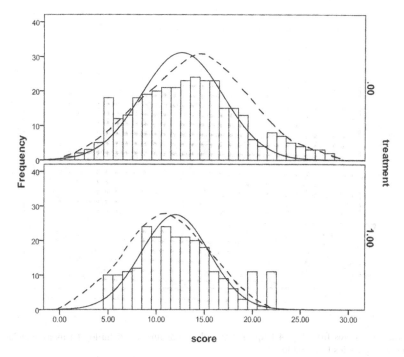

Fig. 4.1 The frequency distributions of the vascular lab scores of untreated (0.00), and treated (1.00) La Fontaine stage I patients. The continuous Gaussian curves are calculated from the mean ± standard deviation, the interrupted Gaussian curves from Bhattacharya modeling

Table 4.1 The frequency distribution of the vascular lab scores of 244 untreated La Fontaine stage I patients (treatment$_0$ patients of Fig. 4.1). The log and delta log terms are respectively log transformations of the frequencies and differences between two subsequent log transformations

Score	Frequency	Log	Delta log
5	10	1.000	
6	10	1.000	0.000
7	11	1.041	0.041
8	12	1.079	0.038
9	24	1.380	0.301
10	21	1.322	−0.058
11	24	1.380	0.058
12	21	1.322	−0.058
13	21	1.322	0.000
14	20	1.301	−0.021
15	18	1.255	−0.046
16	11	1.041	−0.214
17	9	0.954	−0.087
18	6	0.778	−0.176
19	3	0.477	−0.301
20	21	1.322	0.845
21	1	0.000	−1.322
22	21	1.322	−1.322

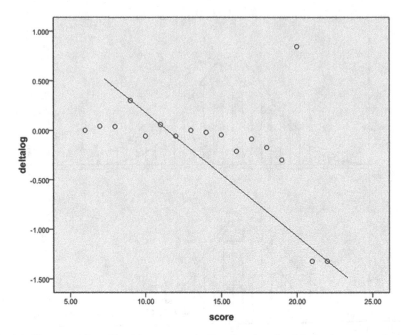

Fig. 4.2 The scores from Fig. 4.1 upper graph plotted against the delta log terms as calculated from the frequencies from Table 4.1

from a representative reference population. Consequently, 2.5 % of the reference Population will exceed the reference range and 2.5 % will be below it. This central 95 %, otherwise called 95 % confidence interval, is calculated from the equation (sd = standard deviation)

$$95\% \text{ confidence interval} \qquad = \text{mean} \pm 1.96 \times \text{sd}$$

$$= 13.236 \pm 1.96 \times 5.600 =$$

$$= \text{between } 2.260 \text{ and } 24.212.$$

Alternatively to the above standard procedure a Bhattacharya procedure can be performed. Table 4.1 shows how it works. We logarithmically transform the frequencies, and then calculate the differences between two subsequent log-frequencies, named the delta log values. Figure 4.2 shows a plot of the scores against these delta log values. A straight line with a correlation coefficient as high as 1.000 is identified, and the equation of this line is used for unmasking the values of the Gaussian curve truly present in these data.

$$y = a + bx$$

$$a = \text{intercept}$$

$$b = \text{direction coefficient}$$

This line is used as the first derivative of a Gaussian curve with

$$\text{mean} = -a/b$$

$$\text{standard deviation} = \sqrt{(-1/b)}.$$

This procedure leads to a result different from that of the standard procedure.

$$
\begin{aligned}
\text{95 \% confidence interval} \quad &= \text{mean} \pm 1.96 \times \text{sd} \\
&= 14.700 \pm 1.96 \times 7.390 = \\
&= \text{between } 0.216 \text{ and } 29.184.
\end{aligned}
$$

The Bhattacharya estimate is wider than the standard estimate, and it is not obvious from the graph which one will best fit the data. Figure 4.1 also shows the graphs of the standard and Bhattacharya Gaussian curves. When counting the tops of the bins cut by either of the curves, it seems that the Bhattacharya curve performs better: 15 cuts versus 9 cuts. And, so, the Bhattacharya 95 % confidence interval produces a better data fit than does the standard 95 % confidence interval as calculated directly from the mean and standard deviation.

4 Improving the p-Values of Data Testing

Figure 4.1 gives an example of frequency distributions of untreated and treated Fontaine stage 1 patients with vascular lab scores on the x-axis and "how often" on the y-axis . We wish to test whether the treatment is better than no treatment. The two sample t-test of these data produced a p-value of 0.051. The non-parametric test of the same data (the Mann–Whitney test) produced a p-value of 0.085. In order to test with improved sensitivity a t-test of the Bhattacharya Gaussian curves is performed. The first two columns of the Tables 4.1 and 4.2 present the x and y axes values of the histograms from Fig. 4.1. First, the y-axis variable is log transformed. Then the differences between two subsequent log transformed y-values are calculated (delta log terms):

$$0.301 - 0 = 0.301$$

$$0.477 - 0.301 = 0.176$$

$$0.699 - 0.477 = 0.222 \text{ etc}$$

A plot of the vascular lab scores against the delta log terms is drawn, and we identify the points that will give you a straight line (Figs. 4.2 and 4.3). A straight line consistent of delta log terms means the presence of a Gaussian distribution in your data. It can be shown that a linear regression analysis of this line serves as the first derivative of a Gaussian curve and that it can be used for calculating the characteristics of this Gaussian distribution in a way that is unaffected by other distributions.

Table 4.2 The frequency distribution of the vascular lab scores of 331 treated La Fontaine stage I patients (treatment$_1$ patients of Fig. 4.1). The log and delta log terms are respectively log transformations of the frequencies and differences between two subsequent log transformations

Score	Frequency	Log	Delta log
1	1	0.000	
2	2	0.301	0.301
3	3	0.477	0.176
4	5	0.699	0.222
5	18	1.255	0.556
6	12	1.079	−0.176
7	13	1.114	0.035
8	18	1.255	0.141
9	19	1.279	0.024
10	20	1.301	0.022
11	21	1.322	0.021
12	21	1.322	0.000
13	23	1.362	0.040
14	24	1.380	0.018
15	23	1.362	−0.018
16	23	1.362	0.000
17	15	1.176	−0.186
18	15	1.176	0.000
19	13	1.114	−0.036
20	6	0.778	−0.336
21	4	0.602	−0.176
22	8	0.903	0.301
23	7	0.845	−0.058
24	5	0.699	−0.146
25	4	0.602	−0.097
26	3	0.477	−0.125
27	3	0.477	0.000
28	2	0.301	−0.176

For the treatment$_0$-data we find

The correlation coefficient R = 0.998
The regression equation is given by y = 0.269 – 0.0183 x
The mean of our Gaussian distribution is given by −0.2690/– 0.0183 = 14.70
De squared standard deviation is given by 1/0.0183 = 54.60
The standard deviation (SD) is $\sqrt{54.60}$ = 7.39.

With n = 331, this would mean that the standard error (SE) of the Gaussian distribution is

$$SE = SD/\sqrt{n} = 0.406.$$

For the treatment$_1$-data we find

The correlation coefficient R = 1.000
The regression equation is given by y = 1.418 – 0.137 x

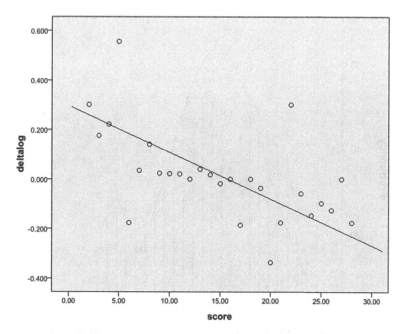

Fig. 4.3 The scores from Fig. 4.1 lower graph plotted against the delta log terms as calculated from the frequencies from Table 4.2

The mean of our Gaussian distribution is given by $-1.418/-0.137 = 10.35$
De squared standard deviation is given by $1/0.137 = 7.299$
The standard deviation (SD) is $\sqrt{7.299} = 2.77$.

With $n = 244$, this would mean that the standard error (SE) of the Gaussian distribution is

$$SE = SD / \sqrt{n} = 0.173$$

An unpaired t-test of these two means produces a t-value of

t-value $= (14.70{-}10.35)/(0.173^2 + 0.406^2) = 9.86$ which means that the two Gaussian curves are largely different with $p = 0.0001$.

5 Objectively Searching Subsets in the Data

Figure 4.4 gives an example of the frequency distributions of vascular lab scores of a population of 787 patients at risk of peripheral vascular disease. Overall normal values of this population can be calculated from the mean and standard deviation:

Normal values = 95 % confidence interval = $24.28 \pm 1.96 \times 11.68$ = between 1.38 and 47.17.

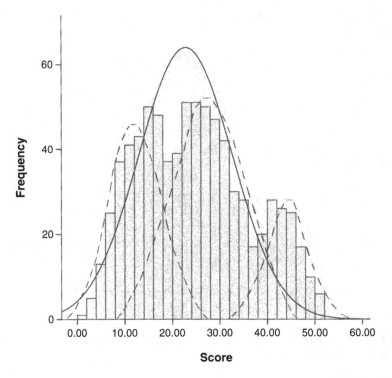

Fig. 4.4 The frequency distributions of vascular lab scores of 787 patients at risk of peripheral vascular disease. The continuous Gaussian curves are calculated from the mean ± standard deviation, the interrupted Gaussian curves from Bhattacharya modeling

The pattern of the histogram is suggestive of certain subsets in this population. Bhattacharya modeling is used for objective searching the subset normal values. Table 4.3 left two columns give the scores and frequencies. The frequencies are log transformed (third column), and, then, the differences between two subsequent log transformed scores are calculated (fourth column). Figure 4.5 shows the plot of the scores against the delta log terms. Three straight lines are identified. Linear regression analyses of these lines produces r-values of 0.980, 0.999, and 0.998.

1. The first regression equation is given by

$$y = 0.944 - 0.078\ x$$

The mean of the corresponding Gaussian curve is given by $-0.944/-0.078$ $= 12.10$.

The squared standard deviation is given by $1/0.078 = 12.82$

The standard deviation (SD) is $\sqrt{12.82} = 3.58$

The normal values of this Gaussian curve is $12.10 \pm 1.96 \times 3.58$, and is between 5.08 and 19.12.

Table 4.3 The frequency
distribution of the vascular
lab scores of 787 patients at
risk of peripheral vascular
disease (the data of Fig. 4.4).
The log and delta log terms
are respectively log
transformations of the
frequencies and differences
between two subsequent log
transformations

Score	Frequency	Log	Delta log
2	1	0.000	0.000
4	5	0.699	0.699
6	13	1.114	0.415
8	25	1.398	0.284
10	37	1.568	0.170
12	41	1.613	0.045
14	43	1.633	0.020
16	50	1.699	−0.018
18	48	1.681	−0.111
20	37	1.570	0.021
22	39	1.591	0.117
24	51	1.708	0.000
26	51	1.708	−0.009
28	50	1.699	−0.027
30	47	1.672	−0.049
32	42	1.623	−0.146
34	30	1.477	−0.176
36	28	1.447	−0.030
38	16	1.204	−0.243
40	20	1.301	0.097
42	28	1.447	0.146
44	26	1.415	−0.032
46	25	1.398	−0.017
48	17	1.230	−0.168
50	10	1.000	−0.230
52	6	0.778	−0.222

2. The second regression equation is given by

$$y = 0.692 - 0.026\,x$$

The mean of the corresponding Gaussian curve is given by $-0.692/-0.026$
$= 26.62$
The squared standard deviation is given by $1/0.026 = 38.46$
The standard deviation (SD) is $\sqrt{38.46} = 6.20$
The normal values of this Gaussian curve is $26.62 \pm 1.96 \times 6.20$, and is between
14.47 and 38.77.

3. The third regression equation is given by

$$y = 2.166 - 0.048\,x$$

The mean of the corresponding Gaussian curve is given by $-2.166/-0.048$
$= 45.13$.
The squared standard deviation is given by $1/0.048 = 20.83$
The standard deviation (SD) is $\sqrt{20.83} = 4.57$
The normal values of this Gaussian curve is $45.13 \pm 1.96 \times 4.57$, and is between
36.17 and 54.09.

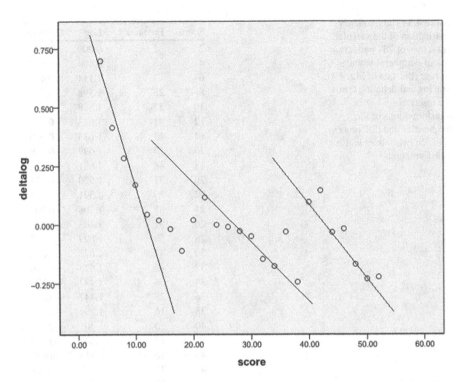

Fig. 4.5 The scores from Fig. 4.4 plotted against the delta log terms as calculated from the frequencies from Table 4.3

In Fig. 4.4 the above three Gaussian curves are drawn as interrupted curves. When there are obviously subsets, to investigators, that is when thing first get very excited. A careful investigation of the potential causes has to be accomplished. The main focus should be on trying to understand any source of heterogeneity in the data. In practice, this may be not be very hard since investigators frequently noticed clinical differences already, and it thus becomes relatively easy to fit the results accordingly. Sometimes differences in age groups or genders are involved. Sometimes also morbidity stages or comorbidities are involved. In the given situation it was decided that the three subsets largely represented (1) stage I la Fontaine patients, (2) patients with risk factors including smoking, and (3) patients with risk factors excluding smoking.

6   Discussion

The current chapter suggests that Bhattacharya modeling not only produces results different from those of the standard approach, but also provides additional benefits. First, it provided a better fit for the data. Second, a better precision as demonstrated by better p-values was obtained. Third, it enabled to better identify certain sub-groups in the data.

The current paper using simulated examples of vascular lab scores has to be confirmed by larger data, but it suggests that Bhattacharya modeling for clinical data analysis tends to perform better than do standards methods.

An important condition for the method to be successful is the presence of Gaussian distributions in the data. In spite of numerous discussions in the literature the conformity of data obtained from patients to a Gaussian distribution is still believed to be of fundamental importance [11–13], and data files including over one million values have been used to confirm this belief [14].

There are, of course, some problems. The first problem is that current clinical research often uses convenience samples from selected hospitals rather than random samples. Particularly, cut-off inclusion criteria like age, gender and laboratory value limits, raises the risk of non-Gaussian data. Goodness of fit tests can be used for checking normality. However, with small samples as commonly observed in clinical research these tests have little power, and a negative goodness of fit test does not exclude the possibility of non-normal data. Also, partly overlapping Gaussian distributions are often present. The method becomes invalid when such distributions are too close to one another, thereby preventing the recognition of the linear part in the first derivative function.

Second, the choice of the appropriate straight lines may sometimes be somewhat subjective: sometimes in a single interval of scores more than a single straight line is possible. E.g., in Fig. 4.2 an almost horizontal line can be drawn though about six delta log terms. However, only lines with a clearly negative direction coefficient are suitable for Bhattacharya modeling.

In spite of the above limitations, we believe that Bhattacharya modeling is a welcome help to clinical data analysis, and we recommend that it be used for the purpose of (1) unmasking normal values of diagnostic tests, (2) improving p-values of testing, and (3) objectively searching subsets in the data. Particularly, when standard data analyses do not produce the expected levels of sensitivity, Bhattacharya modeling is an adequate and more sensitive alternative.

In 1989 Bhattacharya analysis was offered as a Microcomputer Program Package of the Food and Agricultural Organization of the United Nations [2]. However, this program works as a DOS (Disk Operating System) – based application, and is no longer commercially available. Instead, spreadsheets from Excel with Box-Cox transformations (in order to transform the data into the best fit Gaussian distribution) can be conveniently used as explained by Dr. G Jones [15].

7 Conclusions

1. The Bhattacharya method better fitted the data from a single sample than did the usual Gaussian curve derived from the mean and standard deviation with 15 versus 9 cuts.

2. Bhattacharya models demonstrated a significant difference at $p < 0.0001$ between the data from two parallel-groups, while the usual t-test and Mann–Whitney test were insignificant at $p = 0.051$ and 0.085.
3. Bhattacharya modeling of a histogram suggestive of certain subsets identified three Gaussian curves.
4. Bhattacharya analysis can be recommended in clinical research for the purpose of (1) unmasking normal values of diagnostic tests, (2) improving the p-values of data testing, and (3) objectively searching subsets in the data.

References

1. Bhattacharya CG (1967) A simple solution of a distribution into Gaussian components. Biometrics 23:115–135
2. Food and Agricultural Organization of the United Nations. Manuals and Guides BOBP/MAG/14. Separating mixtures of normal distributions: basic programs for Bhattacharya's method and their applications to fish analysis. Copyright@fao.org. 2 Feb 2011
3. Guerin MD, Sikaris KA, Martin A (1992) Pathology informatics: an expanded role for laboratory information systems. Pathology 24:523–529
4. Watson N, Sikaris KA, Morris G, Mitchell DK (1999) Confirmation of age related rise in reference intervals for fasting glucose using the Bhattacharya method and patient data. Clin Biochem Rev 20:92–98
5. Pottel H, Vrydags N, Mahieu B, Vandewynckele E, Croes K, Martens F (2008) Establishing age/sex related serum creatinine reference intervals from hospital laboratory data based on different statistical methods. Clin Chim Acta 396:49–55
6. Baadenhuijsen H, Smit JC (1985) Indirect estimation of clinical chemical reference intervals from total hospital patient data. J Clin Chem Clin Biochem 23:829–839
7. Metz J, Maxwell EL, Levin MD (2002) Changes in folate concentrations following voluntary food fortification in Australia. Med J Austr 176:90–91
8. Zhang L, Liu C, Davis CJ (2004) A mixture model-based approach to the classification of habitats using forest inventory and analysis data. Can J For Res 34:1150–1156
9. Miescke KJ, Musea MN (1994) On mixtures of three normal populations caused by Monogenic inheritance: application to desipramine metabolism. J Psychiat Neurosci 19:295–300
10. Evans DA, Harmer D, Downham DY, Whibley EJ, Idle JR, Ritchie J, Smith RL (1983) The genetic control of sparteine and debrisoquine metabolism in man with new methods of analysing bimodal distributions. J Med Genet 20:321–329
11. Armitage P, Berry G (1994) Statistical methods in medical research. Blackwell Scientific, Oxford, pp 66–71
12. Altman DG (1995) Statistical notes: the normal distribution. BMJ 310:298
13. Feng Z, McLerran D, Grizzle J (1996) A comparison of statistical methods for clustered data analysis with Gaussian error. Stat Med 15:1793–1806
14. Janecki JM (2008) Application of statistical features of the Gaussian distribution hidden in sets of unselected medical laboratory results. Biocyb Biomed Eng 28:71–81
15. Jones G (2006) Reference interval determination by Bhattacharya analysis on skewed distributions- problems and pitfalls. Australasian Association of Clinical Biochemists Annual Scientific Meeting, Hobart, Australia

Chapter 5
Quality-of-Life (QOL) Assessments with Odds Ratios

1 Summary

1.1 Background

The comparison of absolute QOL-scores has been recognized to lack sensitivity to truly estimate QOL.

1.2 Objective

Using an odds ratio approach of QOL scores in 1,350 outpatient clinic patients with stable angina pectoris, the question was assessed that relative scores might provide increased precision to estimate the effects of patient characteristics on QOL data. Increased QOL difficulties were observed in New York Heart Association Angina Class (NYHA) III-IV patients, in patients with comorbidity, as well as in females and elderly patients.

1.3 Methods and Results

Quality of life domains were estimated using a questionnaire based on the Medical Outcomes Short-Form 36 Health Survey and the Angina Pectoris Quality of Life Questionnaire. Results are given as odds ratios = (mean domain scores in patients with characteristic) / (mean domain scores in patients without characteristic).

T.J. Cleophas and A.H. Zwinderman, *Machine Learning in Medicine: Part Two*, DOI 10.1007/978-94-007-6886-4_5,
© Springer Science+Business Media Dordrecht 2013

1.4   Conclusions

Odds ratios can be used in these categories to predict the benefit from treatments. The odds ratio approach of QOL scores provides increased precision to estimate QOL.

2   Introduction, Lack of Sensitivity of QOL-Assessments

Sensitivity defined as ability of the measurement to reflect true changes in QOL (quality of life) is frequently poor in QOL assessments [1]. A well-established problem with QOL scales is their inconsistent relationship between ranges of response and true changes in QOL [2]. A good example of this problem is the physical scale of the SF-36 questionnaire. It ranges from 0 to 100 points. However, while healthy youngsters may score as high as 95 and topsporters even 100, 60 year-old subjects usually score no better than 20. A patient with angina pectoris may score 5 points. If he / she would score 10, instead of 5, after the allowance for sublingual nitrates ad libitum, this improvement would equal 5 % on the absolute scale of 100 points, which does not seem to be very much. However, on a relative scale this score of 10 points is 100 % better than a score of 5 points, and, in terms of improvement of QOL, this difference on the SF-36-scale between 5 and 10 points does mean a world of difference. It, for example, means the difference between a largely dependent and independent way of life. In this example the low score on the absolute-scale masks important and meaningful changes in QOL. The DUMQOL Study Group [3] took issue with this well-recognized but unsolved problem, and performed an odds ratio analysis in a cohort of 1,350 patients with angina pectoris. They showed that this approach provided increased precision to estimate effects of QOL estimators.

3   Real Data Example

The DUMQOL (Dutch Mononitrate Quality of Life) Study Group studied QOL in a cohort of 1,350 patients with stable NYHA (New York Heart Association) angina pectoris, and used the validated form of the Stewart's SF (short form) -36 questionnaire for scoring QOL [4], and the DUMQOL-50 questionnaire for scoring psychological stress and health status [3]. The patients' opinion (patients were requested to estimate the overall amount of his/her QOL as compared to patients they knew with a similar condition) and health status according to the physicians' judgement (the physician was requested to estimate the patients' health status) were scored like the others on 5 point-scales. Internal consistency and retreatment reliability of the test-battery was adequate with Cronbach's alpha 0.66 (otherwise called intraclass correlation, 0.00 = poor, 1.00 = excellent reliability).

4 Odds Ratio Analysis

Table 5.1 gives an overview of the effects of patient characteristics on the QOL estimators in 1,350 patients with stable angina pectoris. Results are presented as odds ratios. The odds ratio presents the relative risk of QOL difficulties and is defined as the ratio between mean domain score of patients with a particular characteristic and that of patients without this particular characteristic.

The procedure readily identifies categories of patients that, obviously, have poor QOL scores. E.g.,

1. Increased QOL-difficulties were observed in patients with advanced New York Heart Association (NYHA) anginal class: the higher the anginal class the larger the risk of mobility difficulties, pain, chest pain, anginal pain, and distress.
2. The risk of mobility difficulties was increased in patients with diabetes mellitus, arrhythmias, and peripheral vascular diseases.
3. Patients using sublingual nitrates (and thus presumably very symptomatic) reported more (severe) mobility difficulties, pain, chest pain, and psychological distress.
4. Female patients reported more (severe) mobility difficulties, pain, anginal pain, and distress than their male counterparts.
5. The risk of mobility difficulties increased with age, but, in contrast, elderly patients reported less pain, anginal pain, and distress.

The above categories of patients are, obviously, very symptomatic and should, therefore, particularly benefit from treatments. The beneficial effects of treatments in patients with particular characteristics can be estimated according to the following procedure:

(1) Odds Ratio$_{\text{active treatment/placebo}}$ = (mean domain score in patients on active treatment)/(mean domain score in patients on placebo).

(2) Odds Ratio$_{\text{characteristic/no characteristic}}$ = (mean domain score in patients with particular characteristic)/(mean domain score in patients without this particular characteristic).

The relative risk of scoring in patients with a particular characteristic, if they used active treatment can be estimated and calculated according to:

(3) Odds Ratio$_{\text{characteristic/no characteristic}}$ $\times$ Odds Ratio$_{\text{active treatment/placebo}}$.

Along this line the odds ratio approach to QOL-assessments can be helpful to estimate the effects of cardiovascular drugs on quality of life in different categories of patients with increased precision.

Table 5.1 Stable angina pectoris: effects of patient characteristics on quality of life estimators. Odds ratios and 95 % confidence intervals are given

	Mobility difficulty	Pain general	Early morning pain	Psychological distress	Chest pain	Patient satisfaction
Gender (females/males)	2.5 (1.8–3.3)[a]	2.0 (1.3–3.0)[a]	1.7 (0.6–4.7)	1.3 (0.9–2.0)	2.1 (1.1–3.9)	0.8 (0.3–1.9)
Age (>68/<86 years)	1.4 (1.2–1.5)[b]	1.0 (0.9–1.1)	0.9 (0.9–1.0)	1.0 (0.9–1.0)	1.0 (0.9–1.0)	1.0 (0.9–1.0)
NYHA (III-and-IV / II-and-I)	5.6 (4.8–6.6)[a]	2.8 (2.1–3.5)[a]	46.8 (26.3–83.1)[a]	4.4 (3.5–5.5)[a]	37.2 (23.4–58.9)[a]	0.6 (0.4–1.1)
Smoking (yes/no)	0.8 (0.5–1.1)	1.3 (0.8–2.1)	12.9 (3.0–56.2)[a]	3.2 (2.0–5.2)[c]	0.5 (0.2–1.2)	5.8 (2.1–15.8)[b]
Cholesterol (yes/no)	0.9 (0.7–1.3)	1.4 (0.3–2.0)	1.3 (0.5–3.4)	1.8 (1.2–2.8)[c]	1.8 (0.9–3.4)	1.1 (0.5–2.6)
Hypertension (yes/no)	0.3 (0.2–0.4)[c]	0.5 (0.3–0.7)[c]	0.7 (0.2–0.9)[c]	0.3 (0.2–0.4)[b]	0.5 (0.3–0.9)[c]	1.7 (0.7–4.1)
Diabetes (yes/no)	2.2 (1.5–3.1)[c]	1.1 (0.6–1.9)	9.1 (3.0–28.2)[a]	2.0 (1.1–3.7)[c]	1.8 (0.7–4.6)	1.1 (0.3–4.2)
Arrhythmias (yes/no)	2.9 (2.0–4.1)[b]	1.3 (0.7–2.1)	3.6 (1.3–10)[c]	3.2 (1.9–5.4)[c]	10.2 (4.5–23.4)[b]	1.2 (0.4–3.7)
PVD (yes/no)	11.0 (7.9–15.1)[a]	2.2 (1.4–3.6)[c]	1.1 (0.7–1.7)	2.6 (1.5–4.5)[c]	1.0 (0.4–2.2)	8.3 (2.7–25.7)[b]
Beta-blockers (yes/no)	0.8 (0.7–0.9)[c]	0.8 (0.5–1.1)	1.7 (0.7–4.0)	0.9 (0.6–1.2)	1.3 (0.7–2.2)	3.2 (1.5–6.9)[b]
Calcium channel blockers (yes/no)	1.5 (1.2–1.9)[c]	1.3 (0.9–1.8)	3.2 (1.5–6.6)[c]	2.0 (1.4–2.9)[c]	6.0 (3.4–13.8)[c]	6.5 (3.0–13.8)[c]
Sublingual nitrates (yes/no)	2.6 (2.1–3.3)[a]	3.0 (2.2–4.2)[a]	1.0 (0.7–1.4)	3.1 (2.5–4.3)[a]	7.1 (4.2–12.0)[a]	3.4 (1.6–6.9)[a]

Quality of life domains were estimated using a questionnaire based on the Medical Outcomes Short-Form 36 Health Survey and the Angina Pectoris Quality of Life Questionnaire. Results are given as odds ratios = (mean domain scores in patients with characteristic) / (mean domain scores in patients without characteristic)

PVD peripheral vascular disease, *NYHA* New York Heart Association Angina Class

[a] P < 0.001
[b] P < 0.01
[c] P < 0.05

5 Discussion

A problem with current QOL-batteries is the inconsistent relationship between ranges of response and true changes in QOL-assessments. This is mainly due to very low (and very high) scores on the absolute-scale, masking important and meaningful changes in QOL. The DUMQOL Study Group showed that this problem can be adequately met by the use of relative rather than absolute scores, and it used for that purpose an odds ratio-approach of QOL scores. This approach provided increased precision to estimate effects on QOL estimators. An additional advantage of the latter approach is that odds ratios are well understood and much in use in the medical community, and that (those) results from QOL research can, therefore, be more easily communicated through odds ratios than through the comparison of absolute scores. For example, "the odds ratio of (severe) mobility difficulties for mononitrate therapy in patients with stable angina is 0.83 ($p < 0.001$)" is better understood than "the mean mobility difficulties score decreased from 1.10 to 1.06 on a scale from 0 to 4 ($p = 0.007$)".

We conclude that recent QOL-research from the DUMQOL Study Group allows for some relevant conclusions, pertinent to both clinical practice and clinical research. QOL should be assessed in a subjective rather than objective way, because the patients' opinion is an important independent contributor to QOL. The comparison of absolute QOL-scores lacks sensitivity to truly estimate QOL. For that purpose the odds ratio approach of QOL scores provides increased precision to estimate QOL.

6 Conclusions

The comparison of absolute QOL-scores has been recognized to lack sensitivity to truly estimate QOL.

Using an odds ratio approach of QOL scores in 1,350 outpatient clinic patients with stable angina pectoris, the question was assessed that relative scores might provide increased precision to estimate the effects of patient characteristics on QOL data. Increased QOL difficulties were observed in New York Heart Association Angina Class (NYHA) III-IV patients, in patients with comorbidity, as well as in females and elderly patients.

Quality of life domains were estimated using a questionnaire based on the Medical Outcomes Short-Form 36 Health Survey and the Angina Pectoris Quality of Life Questionnaire. Results are given as odds ratios = (mean domain scores in patients with characteristic) / (mean domain scores in patients without characteristic).

Odds ratios can be used in these categories to predict the benefit from treatments. The odds ratio approach of QOL scores provides increased precision to estimate QOL.

References

1. Ware JE, Snow KK, Kosinski M, Gandek B (1993) SF-36 Health survey: manual and interpretation guide. The Health Institute, New England Medical Center, Boston
2. Testa MA, Simonson DC (1996) Assessment of quality-of-life outcomes. N Engl J Med 334:835–840
3. Niemeyer MG, Kleinjans HA, De Ree R, Zwinderman AH, Cleophas TJ, Van der Wall EE (1997) Comparison of multiple dose and once-daily nitrate therapy in 1350 patients with stable angina pectoris. Angiology 48:855–863
4. Stewart AL, Hays RD, Ware JE (1988) The MOS short form general health survey. Med Care 26:724–735

Chapter 6
Logistic Regression for Assessing Novel Diagnostic Tests Against Control

1 Summary

1.1 Background

Qualitative diagnostic tests commonly produce false positive and false negative results. Smooth ROC (receiver operated characteristic) curves are used for assessing the performance of a new against a standard test. This method, called c-statistic (concordance statistic) has limitations.

1.2 Objective

This chapter was written to show whether logistic regression with the odds of disease as outcome and the test scores as covariate can be used as an alternative approach. Also, to compare the goodness of either of the two methods.

1.3 Methods

Using as examples vascular lab scores we assessed the performance of logistic regression as compared to c-statistic.

1.4 Results

The c-statistic produced AUCs (areas under the curve) of respectively 0.954 and 0.969 (standard errors 0.007 and 0.005), mean difference 0.015 with a pooled

T.J. Cleophas and A.H. Zwinderman, *Machine Learning in Medicine:*
Part Two, DOI 10.1007/978-94-007-6886-4_6,
© Springer Science+Business Media Dordrecht 2013

standard error of 0.0086. This meant that the new test was not significantly different from the standard test at $p = 0.08$. Logistic regression of these data with presence of disease as dependent and vascular lab scores as independent variable produced regression coefficients of 0.45 and 0.58 with standard errors of respectively 0.04 and 0.05. This meant that the new test was a significantly better predictor of disease than the standard test at $p = 0.04$.

1.5   Conclusions

Logistic regression with presence of disease as dependent and test scores as independent variable is better than c-statistic for assessing qualitative diagnostic tests. This is relevant for future diagnostic research.

2   Introduction

The goodness of novel quantitative diagnostic tests against a standard test can be assessed with linear regression with the diagnostic result as predictor (independent) variable and the severities of disease as outcome (dependent) variable: if the R-square value of the novel test is significantly closer to 1.00 than the standard test, then it is concluded that the novel test performs significantly better than the standard test. However, unfortunately, in clinical research many diagnostic tests have *qualitative* rather than quantitative outcome variables, e.g., a clinical event/disease or not, and linear regression is not applicable for judging the goodness of such tests.

Logistic regression with the odds of disease as outcome (dependent) variable and the test-scores as covariate (independent variable) could be used as a method to model such data. The following reasoning can be used here. If the threshold for a positive test is taken high, then the proportion of false positives will be small. The steeper the logistic regression line, the faster this will happen. In contrast, if the threshold is taken low, then the proportion of false negatives will be small. The steeper the logistic regression line, the faster also this will happen. This would mean that the steeper the logistic regression line, the fewer false positives and false negatives, and thus the better the diagnostic test is. A pleasant aspect of this approach is that absolute instead of relative risks are measured. The current chapter uses as examples vascular lab scores to investigate the performance of logistic regression as compared to c-statistic, an alternative and more traditional approach for the purpose. As examples a non-invasive test for the diagnosis of peripheral vascular disease and a modified version of the test were used in respectively 640 and 587 patients.

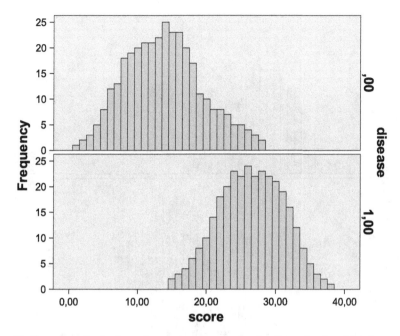

Fig. 6.1 Frequency distributions of a non-invasive test for the diagnosis of peripheral vascular disease (.00 = no disease; 1.00 = disease)

3 Example

Figure 6.1 shows the histograms of vascular lab scores in 640 patients with peripheral vascular disease and healthy controls. The Figure shows that the test is not perfect at all with considerable overlap between the patients with and without vascular disease. Figure 6.2 shows the result of a modified, and, possibly, improved test performed in 587 patient group with similar patient characteristics. Again the test is not perfect, but the patterns of the curves have slightly changed. We can not observe from the Figures which of the two tests is the best one.

4 Comparison of the Tests with Binary Logistic Regression

We use a model similar to the linear regression model used for assessing the goodness of quantitative diagnostic tests. However, because the outcome (dependent) variable is a binary rather than continuous variable, logistic regression instead of linear regression has to be applied.

Again SPSS 17.0 is used [1].

We command: regression.... binary logistic .... Dependent variable disease ....covariate score....ok.

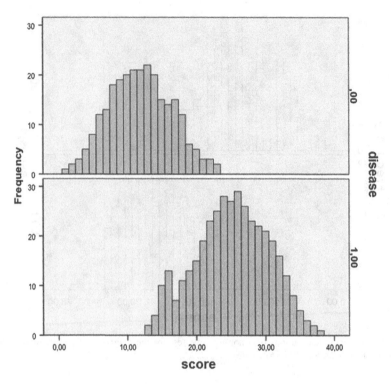

Fig. 6.2 Frequency distributions of a modified version of the test from Fig. 6.1 performed in a age-, sex-, and risk-factor-matched group (.00 = no disease; 1.00 = disease)

The best fit regression equation for test 1 is given underneath:

$$\text{log odds of having the disease} = -9.20 + 0.45 \text{ times the score}$$

The best fit regression equation for test 2 is below:

$$\text{log odds of having the disease} = -9.31 + 0.58 \text{ times the score.}$$

Both regression equations produce highly significant regression coefficients with standard errors of respectively 0.04 and 0.05 and p-values of < 0.0001. The two regression coefficients is tested for significance of difference using the $z-$test:

$$z = (0.58 - 0.45)/\sqrt{(0.04^2 + 0.05^2)} = 0.13/0.064 = 2.03,$$
which corresponds with a p-value of 0.04.

Obviously, test 2 produces a significantly steeper regression model, which means that it is a better predictor of the risk of disease than test 1. We can, additionally, calculate the odds ratios of test 2 versus test 1. The odds of disease with test 1 equals $e^{0.45} = 1.57$, and with test 2 it equals $e^{0.58} = 1.79$. The odds ratio $= 1.79/1.57 = 1.14$, meaning that the second test produces an about 1.14 times better chance of rightly predicting the disease than test 1 does.

5 Comparison of the Tests with Concordance (c)-Statistics

The area under curve of ROC (receiver operated characteristic) curves obtained from the data is often applied as an estimate of the goodness of qualitative diagnostic tests in a way similar to the R-square value for the quantitative diagnostic tests [2, 3]. However, this method, nowadays commonly called the c-statistic (concordance-statistic), has limitations as summarized by Cook [4]: (1) the increase of area under the curve to judge a new (and better) diagnostic test is very small if the standard test already produced a large area under the curve, as commonly observed, and (2) the c-statistic assesses relative risk levels instead of absolute ones, while, in practice, the absolute risk levels are often more important. SPSS 17.0 is used to calculate the c-statistic [1].

We command: Analyze....ROC curve....Test variable score....State variable disease.....value of state variable 1....ROC curve....standard error....OK.

The Figs. 6.3 and 6.4 give the ROC curves of the data from the Figs. 6.1 and 6.2 respectively. The software program produces area under the ROC curve values AUCs) of respectively

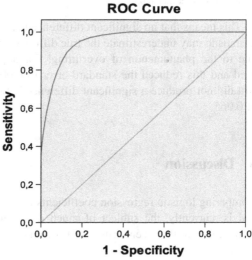

ROC Curve

Diagonal segments are produced by ties.

Fig. 6.3 The space of the area under the curve, otherwise called ROC space, of the data from Fig. 6.1 is calculated by the software to be 0.954 ($= 95.4 \%$)

Fig. 6.4 The space of the area under the curve of the data from Fig. 6.2 is calculated by the software to be 0.969 ($= 96.9$ %)

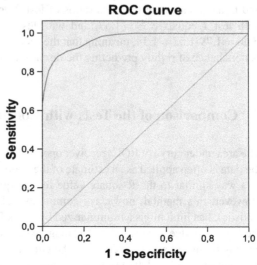

Diagonal segments are produced by ties.

0.954 and 0.969
with standard errors of 0.007 and 0.005.
The pooled standard error equals $\sqrt{(0.007^2 + 0.004^2)} = 0.0086$.
The mean difference of the AUCs $= 0.969{-}0.954 = 0.015$.

The t-test produces a t-value of
$0.015/0.0086 = 1.74$.
This corresponds with a p-value of 0.08,
which is larger than 0.05.

This means that no significant difference between the two tests is demonstrated. C-statistic may underestimate the true difference between two smooth ROC curves due to the phenomenon of overfitting. For adjustment bootstrap sampling was used and this reduced the standard error to 0.0081. The adjusted t-value of 1.85 still did not produce a significant difference between the two tests with a p-value of 0.065.

6   Discussion

Comparing logistic regression coefficients between samples is not straightforward and is, currently, the subject of much discussion in the statistical literature as summarized in the recent article of Karlson et al. [5]. With quantitative diagnostic tests each individual score is associated with a single most probable severity of a disease. With qualitative diagnostic tests, however, each individual score is associated with four different possible outcomes: a false positive or negative outcome, a true positive or negative outcome. The meaning of the b-values of logistic equations is entirely different from those of linear equations. With the linear regression equation $y = bx$ the b-value estimates the ratio y/x. With logistic regression things are much more complex: it estimates the log odds ratio of having a disease versus having no disease. This complex relationship is the main reason, that, so far, logistic regression has not been applied for assessing the performance of qualitative diagnostic tests, and that methods like c-statistic have been developed. The current paper shows that after redefinition of the term performance in terms of "odds ratio of having disease versus having no disease" logistic regression can adequately be used for performance assessments.

The current chapter not only shows that logistic regression can be conveniently applied for assessing the performance of qualitative diagnostic tests, but also suggests that the method is better than c-statistic for that purpose.

Also longitudinal diagnostic tests can be assessed in this way. However, instead of logistic models Cox proportional hazard models are required. Log hazard instead of log odds and the corresponding regression coefficients, the b-values, will have to be used.

Recently, Pencina and D'Agostino [6] proposed the method of reclassification as another alternative to c-statistic. This method uses logistic models like the ones described in this paper to find cut-off test-scores for predicting classes with different risk risks of disease. McNemar's tests are used for comparing partitioned areas under the curves of different data samples. The problem with this method is that, unlike the method reviewed in this paper, it must be applied with paired data samples, which may be hard to obtain with large datasets required for meaningful assessment of small differences between two largely similar diagnostic tests.

Additional advantages of the logistic model compared to c-statistic have to be mentioned:

1. Absolute rather than relative risks of disease are assessed. The c-statistic uses sensitivities and specificities which are relative risks of being truly positive and truly negative, while logistic regression uses the absolute scores as predictor of disease.
2. A limitation of c-statistic is the following. The increase of area under the curve, to judge a new (and better) diagnostic test is very small if the standard test already produced a large area under the curve, as commonly observed.

For performance assessment of quantitative diagnostic tests linear regression is adequate. The current chapter shows that for performance assessment of qualitative diagnostic tests logistic regression is adequate, and seems to provide a better result than does the current c-statistics.

6.1   Conclusions

Logistic regression with presence of disease as outcome and scores as predictor variable is better than c-statistic for the purpose of comparing the performance of qualitative diagnostic tests. This finding is relevant to future diagnostic research.

7   Conclusions

Qualitative diagnostic tests commonly produce false positive and false negative results. Smooth ROC (receiver operated characteristic) curves are used for assessing the performance of a new against a standard test. This method, called c-statistic (concordance statistic) has limitations. This chapter was written to assess whether logistic regression with the odds of disease as outcome and the test scores as covariate can be used as an alternative approach. Also, to compare the goodness of either of the two methods.

Using as examples vascular lab scores we assessed the performance of logistic regression as compared to c-statistic.

The c-statistic produced AUCs (areas under the curve) of respectively 0.954 and 0.969 (standard errors 0.007 and 0.005), mean difference 0.015 with a pooled standard error of 0.0086. This meant that the new test was not significantly different from the standard test at $p = 0.08$. Logistic regression of these data with presence of disease as dependent and vascular lab scores as independent variable produced regression coefficients of 0.45 and 0.58 with standard errors of respectively 0.04 and 0.05. This meant that the new test was a significantly better predictor of disease than the standard test at $p = 0.04$.

We conclude that logistic regression with presence of disease as dependent and test scores as independent variable was better than c-statistic for assessing qualitative diagnostic tests. This may be relevant to future diagnostic research.

References

1. SPSS Statistical Software 17.0 (2010) www.SPSS.com. 20 Oct 2010
2. Green DM, Swets JM (1966) Signal detection theory and psychophysics. Wiley, New York
3. Zou KH, Hall WJ, Shapiro DE (1997) Smooth receiver operated characteristic curves. Stat Med 116:2143–2156
4. Cook NR (2007) Use and misuse of the receiver operating characteristic curve in risk prediction. Circulation 115:928–935
5. Karlson KB, Holm A, Breen R (2011) Comparing regression coefficients between models using logit and probit: a new method. Soc Methodol 42:1–43
6. Pencina MJ, D'Agostino RB, Vasan RS (2008) Evaluating the added predictive ability of a new marker: from area under the ROC curve to reclassification and beyond. Stat Med 27:157–172

Chapter 7
Validating Surrogate Endpoints

1 Summary

1.1 Background

The International Conference of Harmonisation (ICH) Guideline E9, Statistics Principles for Clinical Trials, recommends that surrogate endpoints in clinical trials be validated using either (1) the sensitivity-specificity approach or (2) regression analysis. The problem with (1) is that an overall level of validity is hard to give, and with (2) that a significant correlation between the surrogate and true endpoint is not enough to indicate that the surrogate is a valid predictor.

1.2 Objective

This chapter provides procedures that takes into account the above two problems, and uses, in addition to sensitivity/specificity assessment, an overall validity assessment of the surrogate endpoint test. Also sample size requirement were assessed.

1.3 Methods

Simulated data were used.

T.J. Cleophas and A.H. Zwinderman, *Machine Learning in Medicine: Part Two*, DOI 10.1007/978-94-007-6886-4_7,
© Springer Science+Business Media Dordrecht 2013

1.4   Results

1. In addition to the sensitivity-specificity approach an overall validity level, was used, expressed as the percentage of patients with a true surrogate test, either positive or negative. We calculated confidence intervals of this estimate, and assessed whether they were entirely within the prespecified interval of validity. If so, the surrogate marker was validated for use in subsequent trials.
2. For validating continuous surrogate variables, regression analysis was used, accounting both the correlation between the surrogate and true endpoints, *and* the associations between these two variables and the treatment modalities to be tested. If the proportion of variability in the surrogate endpoint explained the true endpoint by 70 % or more, the surrogate test was validated. A wrong conclusion would be here to accept validity if the surrogate endpoint was an independent determinant of the true endpoint, but not of the treatment modality.
3. In trials using surrogate endpoints the sample size has to be based not only on the expected treatment efficacy but also accounting the validity of the surrogate marker used.

2   Introduction

Clinical trials are often constructed with surrogate endpoints for practical or cost considerations, for example, lipid levels as a surrogate for arteriosclerosis, arrhythmias for coronary artery disease, and cervical smears for tubal infections [1–5]. Such trials make inferences from surrogate observations about the effect of treatments on the supposed true endpoints without accounting the strength of association between the surrogate and true endpoints. The main problem with this practice is that the surrogate endpoint may lack sufficient validity to predict the true endpoint, giving rise to misleading trial results. The International Conference of Harmonisation (ICH) Guideline E9 Statistics Principles for Clinical Trials [6] recommends that, for the approval of a surrogate marker, (1) a statistical relationship with the true endpoint in observational studies be demonstrated, (2) evidence be given from clinical trials that treatment effects on the surrogate correspond to those on the true clinical endpoint, and (3) the surrogate marker like a diagnostic test be tested for sensitivity and specificity to predict the true endpoint. There is, thus, considerable consensus to routinely assess the accuracy of surrogate markers, but not specifically how to do so. Problems with the current sensitivity-specificity approach to validity is, that it is dual and that an overall level of validity is, therefore, hard to give [7]. Also, it can be used for binary (yes/no) endpoints only. As an alternative, regression-models have been proposed [6, 8]. However, a correlation of borderline statistical significance between the surrogate and the true endpoint is not enough to indicate that the surrogate is an accurate predictor. The current chapter underscores the need for accuracy assessment of surrogate endpoints by comparing the required sample sizes of trials with and without surrogate endpoints, and describes two procedures for assessment. The first makes use of an overall level of accuracy with confidence intervals and a prespecified boundary of accuracy. The second uses a regression model that accounts both the association between the surrogate and the true endpoint, and the association between either of these variables and the treatments to be tested.

3 Some Terminology

Surrogate marker/endpoint/test	Laboratory measurement or physical sign used as a substitute for a clinically meaningful endpoint that measure directly how a patient feels, functions, or survives, otherwise called the true endpoint
Validity of a surrogate test	The surrogate test's ability to show which individuals have a true test either positive or negative. We sometimes use the term overall validity to emphasize that the approach is different from assessing sensitivity and specificity separately
Sensitivity	Chance of a true positive surrogate test
Specificity	Chance of a true negative surrogate test
Odds ratio (OR)	Odds of the clinically meaningful endpoint in the treatment group/odds of it in the control group
Alpha (α)	Type I error, chance of finding a difference where there is none
Beta (β)	Type II error, chance of finding no difference where there is one
Null-hypothesis	The study is negative, the treatment does not work. The null-hypothesis of no treatment effect is rejected when the difference from a zero effect is significant
Variance	Estimate of spread or precision in the data. Variance of proportion $p = p(1 - p)$
Standard error (SE)	$\sqrt{}$ (variance/n), where n = sample size
Confidence interval (CI)	It covers a percentage of the results that can be expected if the study would be repeated many times. E.g., 95 % CI between an OR of 1.10 and 1.86 means that 95 % of many similar studies would produce an OR between 1.10 and 1.86
	95 % CI of a proportion be calculated according to: proportion $\pm 1.96*SE_{proportion}$, where * is the sign of multiplication
Prespecified boundary of validity	It is often chosen on clinical grounds, and covers the range of results that are accepted by the investigators as sufficiently valid to use the surrogate test for its purpose. Currently, it is considered good statistical practice to define a prespecified boundary of your expected validity, and, then, test whether the confidence interval of your calculated level of validity falls entirely within the prespecified boundary. If so, you accept, if not you reject the presence of validity
Dependent variable	y-variable in a regression analysis

(continued)

(continued)

Independent variable	x-variable in a regression analysis
Correlation coefficient squared (r^2)	Estimate of strength of association between paired observations. If $r^2 = 0$, there is no association, if $r^2 = 1$, there is 100 % association. If $r^2 = 0.5$, there is 50 % association. One variable determines the other by 50 %, and there is 50 % uncertainty. The r^2-value expresses the proportion of variability in the y-variable determined by the variability in the x-variable
Regression coefficient (b)	Estimate of strength of association between paired observations particularly used in the case of multiple regression

4   Validating Surrogate Markers Using 95 % Confidence Intervals

The validity of a surrogate marker can, like a diagnostic test, be assessed by sensitivity and specificity to predict the true endpoint. In addition to this dual approach to accuracy, an overall validity can be calculated as illustrated below.

		Observed surrogate endpoint (n)	
		Yes	No
Observed true endpoint	Yes	a	b
	No	c	d

Sensitivity $= a/(a + c)$
Specificity $= d/(b + d)$
1-specificity $= b/(b + d)$
Prevalence of true endpoint $= (a + b)/(a + b + c + d)$
The variance of sensitivity is given by $ac/(a + c)^3$.
For the specificity the variance $= db/(d + b)^3$.
Also for 1-specificity the variance $= db/(d + b)^3$
For the prevalence of the true endpoint the variance
$= (a + b)(c + d)/(a + b + c + d)^3$
Overall validity $=$ sensitivity*prevalence $+$ specificity*
(1-prevalence).
$^* =$ the sign of multiplication

For approval of a surrogate marker a boundary of validity is prespecified in the study protocol, e.g., 75 % < validity < 100 %, and confidence intervals of the validity levels are calculated. If the confidence interval falls entirely between the prespecified boundary, validity is demonstrated. E.g., the true endpoint is a cardiovascular event, the surrogate endpoint is an elevated C-reactive protein level, currently a widely used marker for cardiovascular disease.

For the calculation of the confidence intervals of sensitivity and specificity standard errors of these terms are required.

If, e.g.,

$$\text{sensitivity} = 80\,\% \text{ with SE} = 2\,\%,$$

$$\text{specificity} = 90\,\% \text{ with SE} = 1\,\%,$$

then the 95 % confidence intervals of sensitivity and specificity will equal

$$0.80 \pm 1.96 * 0.02 = \text{between } 0.76 \text{ and } 0.84$$

$$0.90 \pm 1.96 * 0.01 = \text{between } 0.88 \text{ and } 0.92.$$

A surrogate endpoint test with a prespecified interval of validity both

$$75\,\% < \text{sensitivity} < 100\,\% \text{ and}$$

$$75\,\% < \text{specificity} < 100\,\%$$

can be validated. The above surrogate endpoint test is valid.

For the calculation of the confidence intervals of the overall validity standard errors (SEs) of this term are equally are required. Calculating the standard error (SE) (= $\sqrt{}$ (variance)) of the overall validity is less straightforward, and has to make use of the formula:

$$\text{Var}(X + Y) = \text{Var}(X) + \text{Var}(Y) + 2\,\text{Cov}(X, Y).$$

$$\text{Var (overall validity)} = \text{Var}(\text{sens} * \text{prev}) + \text{Var}(\text{spec}) * (1 - \text{prev})$$

$$+ 2\text{Cov}\,(\text{sens} * \text{prev}, \ \text{spec} * (1 - \text{prev})).$$

Var = variance;   sens = sensitivity;   spec = specificity;   prev = prevalence; cov = covariance.

The variance of $X * Y$ may be approached from

$$\text{Var}(X * Y) = Y^2\text{Var}(X) + X^2\text{Var}(Y).$$

Using this formula we will end up finding:

$$\text{Var(overall validity)} = \text{prev}^2 * \text{Var}(\text{sens}) + (1 - \text{prev})^2 * \text{Var}(1 - \text{spec})$$

$$+ (\text{sens} - \text{spec})^2 * \text{Var}(\text{prev}).$$

If, e.g.,

$$\text{sensitivity} = 80\,\% \text{ with SE} = 2\,\%,$$

$$\text{specificity} = 90\,\% \text{ with SE} = 1\,\%,$$

$$\text{Prevalence} = 10\,\% \text{ with SE} = 3\,\%,$$

then we can calculate:

$$\text{overall validity} = 0.8 * 0.1 + (0.9) * (1 - 0.1) = 0.89$$

and

$$\text{Var(overall validity)} = 0.1^2 * 0.02^2 + (1 - 0.1)^2 * 0.02^2 + (0.8 - 0.9)^2 * 0.03^2$$
$$= 0.000337.$$

The SE of the overall validity is the square root of the variance, and equals $0.018356 = 1.8356\%$.

This approach makes use of the so-called delta-method which describes the variance of natural logarithm (ln) (X) as $\text{Var(ln(x))} = \text{Var(x)}/x^2$. The approach is sufficiently accurate if the standard errors of prev, sens and spec are small which is true if samples are large.

An overall validity of 89 % with SE 1.8356 % means that the 95 % confidence interval is between

$$0.89 - 1.96 * 0.018356 \text{ and}$$

$$0.89 + 1.96 * 0.018356,$$

and is thus between 85.4 % and 92.6 %. This interval falls entirely between the prespecified interval of validity of 85 % < validity < 100 %. This surrogate endpoint is, thus, validated.

5   Validating Surrogate Endpoints Using Regression Modeling

Table 7.1 shows the total and LDL cholesterol levels being used as tentative surrogate endpoints for coronary artery diameter. For the validation of the two surrogate endpoints the following linear model is used:

$$y = a + b_1 x_1 + b_2 x_2$$

$y = $ true endpoint,
$x_1 = $ treatment modality ($0 = $ placebo, active treatment $= 1$)
$x_2 = $ surrogate endpoint

$$y = a + b_1 x_1$$

r^2 of this equation $= $ proportion variability in y explained by x_1

$$y = a + b_1 x_1 + b_2 x_2$$

r^2 of this equation $= $ proportion variability in y explained by x_1 and x_2.

Table 7.1 Total cholesterol and LDL-cholesterol levels are used as tentative surrogate endpoints for coronary artery diameter

Pt no.	Cor art (mm)	Treat	Tchol (mmol/l)	LDLchol (mmol/l)	Pt no.	Cor art (mm)	Treat	Tchol (mmol/l)	LDLchol (mmol/l)
1	24	0	4.0	2.4	18	12	0	2.0	1.0
2	30	0	6.5	3.2	19	26	0	5.0	2.8
3	25	0	7.5	2.4	20	20	1	4.0	2.0
4	35	1	5.0	3.6	21	43	0	8.0	4.4
5	39	1	4.5	3.8	22	31	0	7.5	3.0
6	30	0	5.0	3.0	23	40	1	7.0	3.8
7	27	0	4.0	2.6	24	31	0	3.5	3.2
8	14	0	2.5	1.6	25	36	1	6.0	3.4
9	39	1	6.5	4.0	26	21	0	3.0	2.0
10	42	1	7.5	4.2	27	44	0	9.5	4.6
11	41	1	5.5	4.0	28	11	1	2.5	1.0
12	38	1	5.5	3.8	29	27	0	4.0	2.6
13	39	1	6.0	3.6	30	24	0	4.5	2.6
14	37	1	5.0	3.4	31	40	1	7.5	3.8
15	47	1	9.0	4.8	32	32	1	3.5	3.4
16	30	0	6.5	2.8	33	10	0	3.0	0.8
17	36	1	6.0	3.8	34	37	1	7.0	3.2
					35	19	0	3.5	2.0

Pt no. patient number, *Cor art* coronary artery diameter, *Treat* treatment modality (0 = placebo. 1 = active treatment), *Tchol* total cholesterol level, *LDLchol* LDL-cholesterol level

Table 7.2 Analysis of associations between true endpoint, treatment modality and surrogate endpoints from Fig. 7.1

	r^2-value	F-value	p-value
True vs treat	0.250	10.9	0.002
LDL-chol vs treat	0.202	8.3	0.007
Tchol vs treat	0.044	1.5	0.226
True vs LDL-chol	0.970	1,052.9	0.000
True vs Tchol	0.630	56.1	0.000

		b-value	Standard error	p-value	r^2
True vs treat and	Treat	0.0135	0.006	0.032	0.98
LDL-chol	LDL-chol	0.891	0.003	0.000	
True vs treat and	Treat	0.375	0.005	0.000	0.75
Tchol	Tchol	0.685	0.018	0.001	

True true endpoint, *treat* treatment modality (0 or 1 for placebo and active treatment), *LDL-chol* surrogate endpoint LDL-cholesterol level, *Tchol* surrogate endpoint total cholesterol level, r^2 Pearson's correlation coefficient squared, *b* regression coefficient

The subtraction sum of the two r^2-values = proportion variability y explained by the surrogate endpoint x_2; the larger the subtraction sum the better the surrogate endpoint. Table 7.2 gives a summary of the calculations. Both LDL-cholesterol and total cholesterol levels are significant predictors of the true endpoint in the multiple regression model with respectively $b = 0.891$, se $= 0.003$, p $= <0.0001$ and $b = 0.685$, se $= 0.018$, p < 0.0001. However, the subtraction sum of the r^2-values is $0.75 - 0.25 = 0.50$ for total cholesterol and $0.98 - 0.25 = 0.73$ for LDL-cholesterol. If the surrogate endpoint is made the dependent variable instead of the true endpoint, then LDL-cholesterol performs better than does total cholesterol. For LDL-cholesterol $r^2 = 0.20$, p-value < 0.01, power approximately 80 %; for total cholesterol $r^2 = 0.04$, p $= 0.226$.

We can conclude that in order to establish a powerful correlation between treatment modality and a surrogate endpoint (p < 0.01, power > 80 %), the proportion variability in y explained by the surrogate endpoint should be close to 70 % or more for accurate predictions.

A wrong method is to accept as a valid result a surrogate endpoint that is a significant determinant of the true endpoint but not of the treatment modality.

We should add that different regression models are more convenient for different data like logistic regression models for odds ratios and Cox regression for survival data, but that the approach, otherwise, is similar.

6   Calculating the Required Sample Size in a Trial with Surrogate Endpoints

The validity or accuracy of a surrogate marker can be expressed in terms of sensitivity and specificity to predict the true endpoint, e.g. healings.

	Healings	Non-healings
New treatment (group 1)	170 (E)	140 (F)
Control treatment (group 2)	190 (G)	230 (H)

odds of healing E/F and G/H,
odds ratio (OR) = E/F/G/H
= (170/140)/(190/230) = 1.47.

Figure 7.1 shows that a true endpoint test for the assessment of the above data has a 95 % confidence interval between 1.09 and 1.99, and that it can reject the null-hypothesis of no difference between the two treatments at P < 0.02. If a surrogate test for the assessment of the same data has a sensitivity of 80 % and specificity of 100 %, then the OR will diminish, because the observed numbers of healings will fall by 20 %, and those of the non-healings will rise correspondingly (OR $= 1.38$; 95 % confidence interval 1.01–1.96, P $= 0.05$). If it has a sensitivity of 80 % and specificity of only 90 %, the OR can be calculated to further fall to 1.31 (95 % confidence interval -0.03 to 1.77, p $= 0.10$), and a significance of difference between the two treatments can no longer be demonstrated (Fig. 7.1). Obviously, with surrogate markers, rapidly, less certainty is provided to estimate the chance of healing or no-healing. In order to maintain a close to true endpoint level of certainty the sample size will have to be increased.

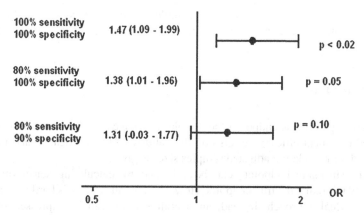

Fig. 7.1 Effect of sensitivity and specificity levels on odds ratios and their 95 % confidence intervals (odds ratio = odds of healing of the new treatment/odds of healing of the control treatment)

The effect on sample size requirement of a reduced sensitivity or specificity is illustrated in the underneath hypothesized example.

In a parallel study group 1 10 % healings are expected,

group 2 20 % healings are expected.

The required sample size can be calculated according to:

$$\text{required sample size} = \text{power index} * \frac{p_1(1 - p_1) + p_2(1 - p_2)}{(p_1 - p_2)^2} \text{ subjects per group}$$

$$= 195 \text{ subjects per group}$$

p_1 = expected proportion of healings in group 1, p_2 = expected proportion of healings in group 2, power index for $\alpha = 0.05$ and $\beta = 0.20$ equals 7.8, * = the sign of multiplication.

If the surrogate test provides 80 % sensitivity, then in group 1 not 10 % but $80 \times 10 \% = 8 \%$ healings will be observed, in group 2 not 20 % but $80 \times 20 \% = 16 \%$. The required sample size will rise to:

$$= 254 \text{ subjects per group.}$$

If sensitivity = 80 % and specificity = 90 %, it can be similarly calculated that the required sample size will further rise to no less than:

$$= 515 \text{ subjects per group.}$$

In trials using surrogate endpoints the sample size has to be based not only on the expected treatment efficacy but also on the validity of the surrogate marker used.

7   Discussion

In trials using surrogate endpoints the sample sizes have to be based not only on the expected treatment efficacy but also on the validity of the surrogate marker used. A method for calculating adjusted samples sizes is given.

Binary surrogate endpoints can be validated by calculating sensitivity and specificity to predict the true endpoint. However, overall validity is hard to quantify using this dual approach. Instead, an overall validity can be expressed by the proportion of patients that have a true surrogate test, either positive or negative, which we called the overall validity level. Still other approaches to the validity of surrogate tests are the so-called positive and negative predictive values and likelihood ratios. Just like the overall validity level, these estimators adjust for numbers of differences in patients with and without the true endpoint, but unlike the overall validity level they do not answer what proportion of patients has a correct test. Riegelman [9], recently, proposed as method for assessing validity of diagnostic tests, the discriminant ability, defined as (sensitivity + specificity)/2. Although this method avoids the dual approach to validity, it wrongly assumes equal importance and equal prevalence of true positive and true negatives, and does neither answer what proportion of the patients has a correct test. We, therefore, decided to use an overall validity level, expressed as the percentage of patients with a true surrogate test, either positive or negative. We calculated confidence intervals of this estimate in order to quantify the level of uncertainty involved in the trial results. If the 95 % confidence interval of the data is entirely within a previously set interval of validity, then the surrogate marker can be validated for use in subsequent trials.

In case of continuous surrogate tests regression models are adequate for testing validity. Not only the association between surrogate and true endpoint must be accounted, but also the associations between either of these variables and the treatment modality to be tested. Interaction assessments are not necessary, if there are no clinical arguments for the presence of interaction. A surrogate test can be validated only, if the proportion of variability in the surrogate endpoint explains the true endpoint by 70 % or more, because the power of the surrogate endpoint to determine the treatment effect is then about 80 %. A wrong conclusion would be to accept adequate validity if the surrogate test is an independent determinant of the true endpoint but not of the treatment modality.

Validating surrogate endpoints can only be done in a trial where a sufficient number of patients reaches both the surrogate and the true endpoint. With mortality or major cardiovascular events as true endpoint large randomized trials with long term follow-up are needed for that purpose. Chen et al. [8] proposed as an alternative a semi-large study with a validation and non-validation set of patients, but this approach is not really different from two separate studies in a single framework. Another interesting alternative was recently proposed by Kassaï et al. They meta-analyzed multiple small studies, but their effort was limited by its post-hoc nature and the heterogeneity of the studies included [10].

If the required sample size or length of follow-up cannot be accomplished, then validity testing of surrogates for true endpoints will be impossible. We will have to look for alternative research methods like looking for intermediate endpoints such as various morbidity measures instead of mortality. We should add that there are additional problems with a true endpoint like mortality: (1) for estimating the effects of preventive medicine that is begun when subjects are middle-aged this endpoint will be statistically weak, because at such ages the background noise of mortality due to other conditions associated with senescence is high, (2) to individual patients low morbidity and high quality of life, generally, means more than does a few additional years of survival. Fortunately, in other research the true endpoint is very well possible, and the surrogate endpoint is pursued because of practical and costs considerations. This applies, e.g., to the example described in the above section. This paper was, particularly, written for the latter purpose. It is to be hoped that the paper will affect the validity of future clinical trials constructed with surrogate endpoints.

8 Conclusions

The International Conference of Harmonisation (ICH) Guideline E9 Statistics Principles for Clinical Trials recommends that surrogate endpoints in clinical trials be validated using either (1) the sensitivity-specificity approach or (2) regression analysis. The problem with (1) is that an overall level of validity is hard to give, and with (2) that a significant correlation between the surrogate and true endpoint is not enough to indicate that the surrogate is a valid predictor. This chapter provides for a nonmathematical readership procedures that avoid the above two problems.

(1) Instead of the sensitivity-specificity approach we used an overall validity level, expressed as the percentage of patients with a true surrogate test, either positive or negative. We calculated confidence intervals of this estimate, and assessed whether they were entirely within the prespecified interval of validity. If so, the surrogate marker was validated for use in subsequent trials. (2) For validating continuous surrogate variables, regression analysis was used, accounting both the correlation between the surrogate and true endpoints, *and* the associations between these two variables and the treatment modalities to be tested. If the proportion of variability in the surrogate endpoint explained the true endpoint by 70 % or more, the surrogate test was validated. A wrong conclusion would be here to accept validity if the surrogate endpoint was an independent determinant of the true endpoint, but not of the treatment modality. It is to be hoped that this chapter will affect the validity of future clinical trials constructed with surrogate endpoints.

References

1. Pratt CM, Moye LA (1995) The cardiac arrhythmias suppression trial. Circulation 91:245–247
2. Canner PL, Berg KG, Wenger NK, Stamler J, Friedman L, Prineas RJ, Friedewald F (1986) Fifteen year mortality of the coronary drug project. J Am Coll Cardiol 8:1245–1255
3. Riggs P et al (1990) Osteoporosis in postmenopausal women. N Engl J Med 32:802–809
4. Fleming TR, DeMets DL (1996) Surrogate end points in clinical trials: are we being misled? Ann Intern Med 125:605–613
5. Boissel JP, Collet HC (1992) Surrogate endpoints: a basis for a rational approach. Eur J Clin Pharmacol 43:235–244
6. Philips A, Haudiquet V (2003) The international conference of harmonisation (ICH) guideline E9, statistics principles for clinical trials. Stat Med 22:1–11
7. Cleophas TJ (2005) Clinical trials: a new method for assessing accuracy of diagnostic tests. Clin Res Regul Aff 22:93–101
8. Chen SX, Leung DH, Qin J (2003) Information recovery in a study with surrogate endpoints. J Am Stat Assoc 10:7–18
9. Riegelman RK (2005) Studying a study and testing a test. Lippincott Williams & Wilkins, Philadelphia
10. Kassaï B, Shah NR, Leizorovicz A, Cucherat M, Gueyffier F, Boissel JP (2005) The true treatment benefit is unpredictable in clinical trials using surrogate outcome measures with diagnostic tests. J Clin Epidemiol 58:1042–1051

Chapter 8
Two-Dimensional Clustering

1 Summary

1.1 Background

In clinical data subgroups can sometimes be identified using regression analysis of subgroup characteristics against some outcome variable, but in data samples without an available outcome variable cluster analysis is a suitable alternative. It is based on the concept that patients with closely related characteristics may also be more related in other fields like prognoses and treatment efficacies.

1.2 Objective

To compare the performance of three different cluster methodologies, hierarchical, k-means, and density-based clustering.

1.3 Methods

A simulated data example of 50 patients with mental depression was used.

1.4 Results

Each cluster methodology identified three clusters. However, the cluster patterns were very different. The hierarchical method showed round patterns different in size,

T.J. Cleophas and A.H. Zwinderman, *Machine Learning in Medicine:*
Part Two, DOI 10.1007/978-94-007-6886-4_8,
© Springer Science+Business Media Dordrecht 2013

the k-means method round patterns equal in size, and the density-based method non-circular patterns also different in size. The patterns from the hierarchical method were better in agreement with the patterns as clinically expected, than those from the other methods.

1.5   Conclusions

1. Cluster analysis is little used in clinical research.
2. Hierarchical cluster is adequate if subgroups in the data are expected to be different in size but, otherwise, Gaussian-like. It is available in the module Classify of SPSS.
3. K-means cluster analysis is adequate if subgroups are expected to be approximately similar in size. It is also available in the module Classify of SPSS.
4. Density-based cluster analysis is adequate if small outlier groups between an, otherwise, homogeneous population is expected. It is not available in SPSS, but an interactive JAVA Applet is freely obtainable at the Internet.

2   Introduction

Populations with a single clinical diagnosis are, otherwise, often very heterogeneous. This may have consequences regarding prognosis and treatment efficacies. E.g., patients with mental depression may suffer from subtypes like reactive depression, depression with insomnia or true depression [1]. Patients with gastric cancer may have different expression levels of genes that are related with their prognoses [2]. Different characteristics in a population of HIV patients were held responsible for their HIV vaccine efficacy [3]. Underlying mechanisms were established for explaining high anti-trypanosomal drug efficacy [4]. Many more examples can be given. Subgroups can sometimes be identified using regression analysis of potential subgroup properties against some outcome variable, but in data samples without an available outcome variable cluster analysis is a suitable alternative. It is based on the concept that patients with closely related characteristics might also be more related in other fields like prognoses and drug efficacies.

Unlike regression analysis, cluster analysis does not require a dependent (outcome) variable. In a sense the patients themselves are the dependent variables. Cluster analysis is currently an important methodology in explorative data mining, and a main task in machine learning, and is sometimes called unsupervised machine learning, because there is, generally, no dependent variable [5]. It is widely used by econometrists and sociologists for identifying population subgroups [6], but in clinical research it is virtually unused. Apart from its current key role in sequence-clustering [7, 8], which is a method for clustering related DNA and protein sequences, we found only sporadic published papers of cluster analysis in any type of health research. Two adverse event studies [9, 10], one drug manufacturing study [11], and one patient compliance study [12] have been published.

The current chapter using a simulated example assesses the potential of cluster analysis for the analysis of clinical data and compares the clustering results of different methodologies, including hierarchical, k-means, and density-based clustering.

3 Three Clustering Methodologies

Three methodologies are currently used.

1. Hierarchical cluster analysis

   It was invented by Robert Sibson (1973), statistician from King's College Cambridge UK statistical department [13] and Daniel Defays (1977), psychologist from Liege University Belgium [14]. A cluster is estimated by the distances between the values needed to connect the cases. The smaller the distance, the more similar the cases are. The distance is calculated as the squared difference between two cases. The method starts with all patients being a cluster of his/her own. Then, the smallest distances are used to form the first clusters. This procedure continues, and stops when all patients are in a cluster. With Gaussian-like data as commonly observed in scientific research, the clusters tend to have an oval pattern and with similarly sized scales even a round pattern, but they are not equal in size.

2. K-means cluster analysis

   It was invented by Stuart Lloyd, a physicist from New Jersey who worked at the Math Department of Bell Telephone in 1957, but was first published in 1982 [15]. Compared to hierarchical clustering this method works in the opposite direction, but, otherwise, largely similar. It does not start with all patients being a cluster of his/her own, but instead, randomly selects cluster centers, and, then by iteration tries and finds the best fit centers for the data given, i.e., those with the shortest distances to the centers. Intuitively one may assume that this procedure should lead to the same result. However, this is not necessarily true. The point is that one important assumption of the k-means method is that the clusters are equally sized, and this is not an assumption of the hierarchical method.

3. Density-based cluster analysis

   It was invented by Martin Ester and Hans Kriegel, professors of computer science at Muenich University in 1996 [16]. Density-based clusters are defined as areas of higher density than the remainder of the data. Individuals in the sparse area are considered as noise (random effects). Density-based clustering connects points that satisfy a density criterion given by a minimum number of patients within a defined radius. Unlike in the above two methods, here the clusters do not need to be round, but they are multiform areas that are, simply, more dense than the cluster-less areas.

4   Example

Fifty patients with mental depression are assessed for age and depression score (0 = very mild, 10 = severest). We will use various cluster methods in order to identify clusters with different ages and severities. We have some prior idea about differences in age and severity between patients with true depression, reactive depression, and depression with insomnia. Table 8.1 gives the patient data.

4.1   Hierarchical Cluster Analysis

Patients are called cases. The distances between the cases are calculated as the squared differences between two cases. We will use SPSS statistical software [17].

> Command: Analyze . . . .Classify . . . .Hierarchical Cluster Analysis . . . .enter variables . . . .Label Case by: case variable with the values 1–50 . . . .Plots: mark Dendrogram . . . .Method . . . .Cluster Method: Between-group linkage . . . .Measure: Squared Euclidean Distance . . . .OK.

Figure 8.1 shows a dendrogram from the data from Table 8.1. The actual distances between the cases are rescaled to fall into a range of 0–25 units (0 = minimal distance, 25 = maximal distance). The cases no. 1–11, 21–25 are clustered together in cluster 1, the cases 12, 13, 20, 26, 27, 31, 32, 35, 40 in cluster 2, both at a rescaled distance from 0 units at approximately 3 units (Fig. 8.2). The remainder of the cases is clustered at approximately 6 units. Obviously, three clusters of cases have been indentified with cases more similar to one another than to the cases of the other clusters. When minimizing the output file, the data file comes up and it now shows the cluster membership of each case. We use SPSS to draw a Dotter graph of the data.

> Command: Analyze . . . .Graphs . . . .Legacy Dialogs: click Simple Scatter . . . .Define . . . .Y-axis: enter Depression Score . . . .X-axis: enter Age . . . .OK.

The graph produced by SPSS is given in triplicate in Fig. 8.1, and the memberships of the cases per method is shown. The upper graph, the hierarchical model, shows that all of the clusters are oval and even approximately round because variables have similarly sized scales, but they are different in size. Two large clusters are in both the youngsters and the elderly, one small cluster is in between. This hierarchical cluster model is in agreement with the patterns as clinically expected: two large populations with respectively true depression (younger patients) and reactive depression (elderly), and one small population with depression associated with insomnia. The method does not provide a statistic to test whether the three clusters are significantly different from one another, but the graph shows that there is a complete separation between the three clusters, and, so, the between-cluster differences must be statistically very significant. No statistical test is needed.

Table 8.1 Data file of the example used, patients are called cases, the cluster membership of the hierarchical and k-means clustering methods are given

Age	Depression score	Patient number	Hierarchical clustering	k-means clustering
20.00	8.00	1	1	1
21.00	7.00	2	1	1
23.00	9.00	3	1	1
24.00	10.00	4	1	1
25.00	8.00	5	1	1
26.00	9.00	6	1	1
27.00	7.00	7	1	1
28.00	8.00	8	1	1
24.00	9.00	9	1	1
32.00	9.00	10	1	1
30.00	1.00	11	1	1
40.00	2.00	12	2	2
50.00	3.00	13	2	2
60.00	1.00	14	3	2
70.00	2.00	15	3	3
76.00	3.00	16	3	3
65.00	2.00	17	3	3
54.00	3.00	18	3	2
54.00	4.00	19	3	2
49.00	3.00	20	2	2
30.00	4.00	21	1	1
25.00	5.00	22	1	1
24.00	4.00	23	1	1
27.00	5.00	24	1	1
35.00	6.00	25	1	1
45.00	5.00	26	2	2
45.00	6.00	27	2	2
67.00	7.00	28	3	3
80.00	6.00	29	3	3
80.00	5.00	30	3	3
40.00	1.00	31	2	2
50.00	2.00	33	3	2
80.00	4.00	34	3	3
50.00	5.00	35	2	2
76.00	6.00	36	3	3
65.00	7.00	37	3	3
79.00	8.00	38	3	3
57.00	3.00	39	3	2
46.00	4.00	40	2	2
54.00	5.00	41	3	2
74.00	6.00	42	3	3
65.00	7.00	43	3	3
57.00	9.00	44	3	2
68.00	8.00	45	3	3

(continued)

Table 8.1 (continued)

Age	Depression score	Patient number	Hierarchical clustering	k-means clustering
67.00	7.00	46	3	3
65.00	6.00	47	3	3
64.00	5.00	48	3	3
74.00	4.00	49	3	3
75.00	3.00	50	3	3

Fig. 8.1 Dendrogram of the 50 cases of Table 8.1. The actual distances between the cases are rescaled to fall into a range of 0–25 units (0 = minimal distance, 25 = maximal distance). The cases 1–11, 21–25 are clustered together in cluster 1, the cases 12, 13, 20, 26, 27, 31, 32, 35, 40 in cluster 2, both at a rescaled distance from 0 at 3. The remainder of the cases are clustered at a distance of 6. At that point, three clusters of cases have been indentified with cases more similar to one another than to the cases of the other clusters. Beyond the distance of 10 only two clusters are left in the data

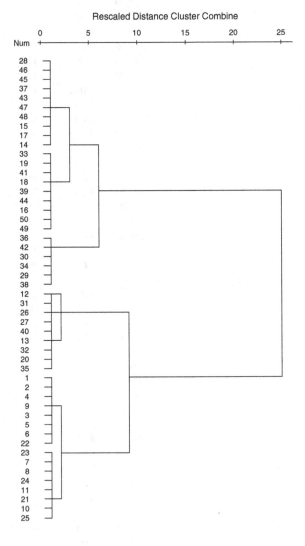

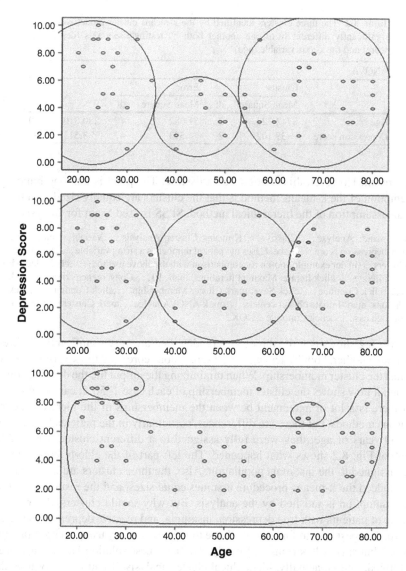

Fig. 8.2 Graphs of the data from the example in this paper: *upper graph* hierarchical cluster analysis, *middle graph* k-means cluster analysis, *lower graph* density-based cluster analysis

4.2 K-means Cluster Analysis

Compared to hierarchical clustering this method works in the opposite direction. It does not start with all patients being a cluster of his/her own, but instead, randomly selects cluster centers, and, then by iteration tries and finds the best fit centers for the data given. Intuitively one may assume that this procedure should lead to the

Table 8.2 The three clusters identified by the k-means cluster model were very significantly different from one another both by testing the y-axis (depression score) and the x-axis variable (age)

ANOVA

	Cluster		Error			
	Mean Square	df	Mean Square	df	F	Sig.
Age	8712.723	2	31.082	47	280.310	.000
Depression score	39.102	2	4.593	47	8.513	.001

same result. However, this is not necessarily true. The point is that one important assumption of the k-means method is that the clusters are equally sized, and this is not an assumption of the hierarchical method. SPSS is used again for analysis.

Command: Analyze....Classify....K-means Cluster Analysis....Variables: enter Age and Depression score....Label Cases by: patient number as a string variable....Number of clusters: 3 (in our example chosen for comparison with the above method)....click Method: mark Iterate....click Iterate: Maximal Iterations: mark 10....Convergence criterion: mark 0....click Continue....click Save: mark Cluster Membership....click Continue....click Options: mark Initiate cluster centers....mark ANOVA table....mark Cluster information for each case....click Continue....OK.

Table 8.2 shows that the three clusters identified are very significantly different from one another, both by testing the y (depression score) and the x variables (age) against the cluster membership. When minimizing the output file the data file comes up, and it now shows the cluster membership of each case 1–50. It can be observed that there is a lot of agreement between the memberships of the hierarchical and k-means methods, but there are differences, particularly in the patients between 50 and 60 years of age: they were fully assigned to a different cluster. The middle graph of Fig. 8.2 shows what happened. The left part of the elderly population is now assigned to the insomnia population. Also, the three clusters are now equal in size. Indeed the k-means procedure assumes equal sizes, and the result shows that this assumption is satisfied by the analysis. But why should clusters of a random sample of patients with true depression, insomnia, and reactive depression be equal in size. The best way to find out would be to repeat the study using a larger random sample, but this is laborious and costly. The next best solution has already been performed, and is, actually, hierarchical cluster analysis, because it only uses the neighborhood criterion and skips the equal size criterion.

4.3   Density-Based Clustering

The DBSCAN method was used (density based spatial clustering of application with noise) [17]. As this method is not available in SPSS, an interactive JAVA Applet freely available at the Internet was used [18]. The DBSCAN connects points that satisfy a density criterion given by a minimum number of patients within a radius given (radius = Eps; minimum number = Min pts).

> Command: User Define . . . .Choose data set: remove values given . . . .enter you own x and y values . . . .Choose algorithm: select DBSCAN . . . .Eps: mark 25 . . . .Min pts: mark 3 . . . .Start . . . .Show.

Three clusters are shown (Fig. 8.2 bottom graph). Two very small ones, one with very high depression scores in youngsters and one with very high depression scores in patients 60–70 years are observed, and one large one with moderate to low levels of depression at all ages. All of the clusters identified are non-circular and, are, obviously, based on differences in patient-density.

5 Discussion

Cluster analysis is, currently, an important methodology in explorative data mining, and a main task in machine learning [4–6], but, unfortunately, little used in health research, in spite of the omnipresence of heterogeneities in patient diagnosis groups. The little use in pharmaceutical research is probably due to the traditional belief of pharmaceutical investigators in clinical trials, where randomization takes care that heterogeneities in the data are equally distributed between the treatment and control groups, and where they are no further taken into account. Controlled clinical trials may, indeed, be more accurate and reliable for making health predictions, but are uncontrolled data completely meaningless? Even if heterogeneities established are clinically relevant in less than 10 % of the cases, 10 % is better than 0 %. Also, 10 % is a lot, if you consider the ready availability of large and complex data files in electronic health records of modern health facilities and other institutions.

In the example of this chapter we have argued that hierarchical clustering was the best way for assessing the data, because of a prior belief, that the clusters were likely to be different in size, and, because we had no arguments for non-Gaussian patterns in these data. However, the other two methods may be more appropriate with other types of data. For example, the k-means method might be more appropriate with clusters having the same size like clusters of genders in a random population. Density-based clustering may be more appropriate with large homogeneous populations and one or more relatively small outlier subsets, like patients with specific environmental, genetic, life style characteristics etc.

Only two-dimensional clusters are reviewed here, with age and depression severity as the only variables. However, for all of the three models reviewed in this paper multidimensional clustering is possible, if you wished to include more than two variables. Multidimensional clustering is relevant in clinical research considering the multifactorial nature of disease and drug efficacy [1–6], and can be performed even if outcome variables are not available.

We should add that two-dimensional density-based clustering [19] and, in some studies, also two-dimensional k-means clustering [20] were of great importance in the field of imaging, like image compression and image color quantization [6]. Unfortunately, it is little used in medical imaging like PET (positive emission tomography) and MRI (magnetic resonance image) scanning [6], but this is a matter of time, now that it is increasingly available in SPSS and other statistical software programs.

6 Conclusions

1. Cluster analysis is little used in clinical research.
2. Hierarchical cluster is adequate if subgroups in the data are expected to be different in size but, otherwise, Gaussian-like. It is available in the module Classify of SPSS.
3. K-means cluster analysis is adequate if subgroups are expected to be approximately similar in size. It is also available in the module Classify of SPSS.
4. Density-based cluster analysis is adequate if small outlier groups between an, otherwise, homogeneous population is expected. It is not available in SPSS, but an interactive JAVA Applet is freely available at the Internet.

References

1. Anonymous (1978) Diagnostic and statistical manual of mental disorders. American Psychiatric Society, New York
2. Solyanik GL (2010) Multifactorial nature of tumor drug resistance. Exp Oncol 32:181–185
3. National Institutes of Health, 9000 Rockville Pike, Bethesda, Maryland (2003) Enhancing HIV vaccine efficacy in high-risk drug users. Release data Jan 6, RFA Number DA-03-002
4. Alsford S, Eckert S, Baker N, Glover L, Sanchez-Flores A, Leubg KF, Turner DJ, Field MC, Berriman M, Horn D (2012) High throughput decoding of antitrypanosomal drug efficacy and resistance. Nature. doi:10.1038/nature_10771
5. Anonymous (2012) Machine learning. http://en.wikipedia.org/Machine_learning. 12 July 2012
6. Anonymous (2012) Cluster analysis. http://en.wikipedia.org/Cluster_analysis. 12 July 2012
7. Anonymous (2012) Sequence clustering. http://en.wikipedia.org/wiki/Sequence_analysis. 12 July 2012
8. Kim HK, Choi IJ, Kim HS, Oshima A, Michalowski A, Green JE (2011) A gene expression signature of acquired chemoresistance to cisplatinum and fluorouracil combination chemotherapy in gastric cancer patients. PLoS One 18:e16694
9. Yeh ST (2012) Clinical adverse event data analysis and visualization. Smith Kline Datafile, 10 July 2012
10. Bychowiec B, Piskorski J, Stanislawska K, Dziarmaga M, Mineczykowski A, Wykretowicz A, Wysocki H (2010) An exploratory clustering study of rare adverse events in drug deluting stent patients. Comput Methods Sci Technol 16:5–11
11. Xu D, Redamn-Furey N (2007) Statistical cluster analysis of pharmaceutical solvents. Int J Pharm 339:175–188
12. Hawwa A, Millership JS, Collier PS, McCarthy A, Dempsey S, Cairns C, McElnay JC (2009) Development of objective methodology to measure medication adherence to oral thiopurines in paediatric patients with acute lymphoblastic leucemia. Eur J Clin Pharmacol 65:1105–1112

13. Sibson R (1973) An optimally efficient algorithm for a single-link cluster method. Comput J 16:30–34
14. Defays D (1977) An efficient algorithm for a complete link cluster method. Comput J 20:364–366
15. Lloyd SP (1982) Least square quantization in PCM. IEEE Trans Inf Theory 28:129–137
16. Ester M, Kriegel HP, Sander J, Xu X (1996) A density based algorithm for discovering cluster in large spatial databases with noise. In: Proceedings of 2nd international conference on knowledge discovery and data mining. AAI Press, Portland
17. SPSS statistical software (2012) www.spss.com. 12 July 2012
18. Data Clustering Applets (2012) http://webdocs.cs.ualberts.ca/~yaling/Cluster/applet. 17 Sept 2012
19. Anonymous (2012) Density-based cluster and outlier analysis. www.dbs.informatik.uni-muenchen.de/Forschung/KDD/Clustering. 25 Sept 2012
20. Kanungo T, Mount D, Netanyahu N, Piatko C, Wu A (2002) An efficient k-means clustering algorithm: analysis and implementation. IEEE Trans Pattern Anal Mach Intell 24:881–892

Chapter 9
Multidimensional Clustering

1 Summary

1.1 Background

In groups with a single diagnosis the individual patient characteristics are often very heterogeneous, and this may affect prognoses and drug efficacies.

1.2 Objective

To assess the potential of multidimensional cluster analysis for the analysis of heterogeneous clinical data

1.3 Methods

Using a simulated example of 30 mentally depressed patients with 9 item scores applied as predictor variables, different clustering methodologies available in SPSS' statistical software were compared, (1) hierarchical, (2) k-means, and (3) two-step clustering.

1.4 Results

All of the methods identified three clusters with identical memberships. Unlike hierarchical clustering, k-means clustering provided test statistics, demonstrating

T.J. Cleophas and A.H. Zwinderman, *Machine Learning in Medicine:*
Part Two, DOI 10.1007/978-94-007-6886-4_9,
© Springer Science+Business Media Dordrecht 2013

that all of the item scores were very significant predictors of the cluster memberships ($p < 0.0001$). Two-step clustering assessed small random subgroups of patients instead of individual patients, and was, therefore, theoretically less computationally intensive. Also the cluster composition and the importance of the separate items for the formation of the clusters were given.

1.5   Conclusions

1. Hierarchical, k-means, and two-step cluster analysis can be adequately used for the analysis of heterogeneous clinical data and can provide relevant prognostic patterns.
2. On a theoretical base two-step analysis is better suitable for large data files.
3. The clustering methods as reviewed may be less accurate and reliable than controlled clinical trials, but, on the other hand, they are more flexible, because they do not require a control group or prior randomization of the data.
4. Also, they may, sometimes, be more adequate than multiple regression analysis, because the presence of collinearity (two explanatory variables that are highly correlated with one another) is much less a problem, and the presence of an outcome variable is not required.

2   Introduction

Populations with a single clinical diagnosis are, otherwise, often very hetero-geneous. This may have consequences regarding prognosis and drug efficacies. E.g., patients with mental depression may suffer from subtypes like reactive depression, depression with insomnia or true depression [1]. Patients with gastric cancer may have different expression levels of genes that are related with their prognoses [2]. Different patient characteristics in a population of HIV patients were held responsible for their HIV vaccine efficacy [3]. Underlying mechanisms were established for explaining high anti-trypanosomal drug efficacy [4]. Many more examples can be given. Subgroups can sometimes be identified using regression analysis of potential subgroup properties against some outcome variable, but in data samples without an available outcome variable cluster analysis is a suitable alternative. It is based on the concept that patients with closely related characteristics might also be more related in other fields like prognoses and drug efficacies.

Unlike regression analysis, cluster analysis does not require a dependent (outcome) variable. In a sense the patients themselves are the dependent variables. Cluster analysis is currently an important methodology in explorative data mining, and a main task in machine learning, and is sometimes called unsupervised machine learning, because there is, generally, no dependent variable [5]. It is widely used by econometrists and sociologists for identifying population subgroups [6], but in pharmaceutical research it is virtually unused. Apart from its current key role in sequence-clustering [7, 8], which is a method for clustering related DNA and protein sequences, we found only sporadic published papers of cluster analysis in treatment efficacy studies. Two adverse event studies [9, 10], have been published, respectively one study of the effects of subsets of genes on chemo-resistance to platinum and fluorouracil, and one general study of frequently observed drug adverse effects. One drug manufacturing study of the effects of 57 pharmaceutical solvents on crystallization optimization [11], and one patient compliance study of adherence factors to oral thiopurines have been published [12].

The current chapter using a simulated example assesses the potential of multidimensional cluster analysis for the analysis of heterogeneous clinical data and compares the clustering results of different methodologies, including hierarchical, k-means, and two-step clustering.

All of the methods are readily available in the Classify module of SPSS statistical software. For the convenience of investigators new to the methods, we will describe the step by step analyses with each of the methods.

3   Three Clustering Methodologies, as Available in SPSS Statistical Software

Three methodologies are currently used.

1. Hierarchical cluster analysis
   It was invented by Robert Sibson (1973), statistician from King's College Cambridge UK statistical department [13] and Daniel Defays (1977), psychologist from Liege University Belgium [14]. A cluster is estimated by the distances between the values needed to connect the cases. Patients are often called cases by the software program. The smaller the distance, the more similar the cases are. The distance is calculated as the squared difference between two cases. The method starts with all patients being a cluster of his/her own. Then, the smallest distances are used to form the first clusters. This procedure continues, and stops when all patients are in a cluster.

2. K-means cluster analysis
   It was invented by Stuart Lloyd, a physicist from New Jersey who worked at the Math Department of Bell Telephone in 1957, but was first published in 1982 [15]. Compared to hierarchical clustering this method works in the opposite direction, but, otherwise, largely similar. It does not start with all patients being a cluster of his/her own, but instead, randomly selects cluster centers, and, then by iteration tries and finds the best fit centers for the data given, i.e., those with the shortest distances to the centers. Intuitively one may assume that this procedure should lead to the same result. However, this is not necessarily true. The point is that one important assumption of the k-means method is that the clusters are equally sized, and this is not an assumption of the hierarchical method.

3. Two-Step Clustering

Two-step clustering is relatively young, and was invented in 1993 by Jeffrey Banfield and Adrian Raftery [16], statisticians at respectively Montana State University in Bozeman, and University of Washington in Seattle. It was, originally, called model-based Gaussian clustering by its inventors, because unlike the other methods it assumes a Gaussian relationship between the number of clusters and the uncertainty of the model. It is a two-step procedure, that starts with autoclustering, i.e., the random formation of a large number of preclusters of individuals that are closest to one another. Then, as a second step, a standard hierarchical procedure is followed, not of the individuals, but of the preclusters formed. Finally, formal goodness of fit statistics for various cluster numbers are computed, which enable the investigator to observe at which cluster number the within cluster distances start being adequately small for his/her purposes. The largest advantage of two-step clustering is that it uses preclusters selected by the software program, instead of individuals. It, therefore, works much faster than the other two methods, that, with large samples, may require weeks or months of computation time.

4   Example

A study of 30 mentally depressed patients was used with 9 item scores applied as predictor variables of severity and prognosis. Table 9.1 gives the data with each row being a patient, and the nine left side columns being the item scores.

4.1   Hierarchical Clustering

A cluster is estimated by the maximum distances between the values needed to connect the cases (patients). The smaller the distance, the more similar the cases are. The distance is calculated by the add-up sum of squared difference between two cases. Table 9.2 explains the computations made with the patients no. 1 and 2 (from Table 9.1). The add-up sum of squared differences between the patients 1 and 2 is given as well as the add-up sum of the squared distances between the two patients, as computed for all items. The values estimate the distance between the two patients. The same procedure is performed for all of the patients, and the smallest distances are, then, used to form the first clusters. Then, at larger distances new clusters are formed by merging the former clusters with additional patients. For that purpose the distances are averaged. This stops, when all patients are in a cluster. We will use SPSS statistical software [17].

> Command: Analyze....Hierarchical Clustering...Variables: enter item scores 1-9....Label Case by: enter case variable with the values 1-30...Statistics....Plots: mark Dendrogram ....Method .... Cluster Method: Between-group linkage....Measure: Squared Euclidean Distance....OK.

Table 9.1 The study file of 30 depressed patients (each patient is given a row) with 9 item scores as predictors of severity and prognosis. In the three right-end columns the cluster results of hierarchical clustering (hier), k-means clustering (k-mean), and two-step clustering (two) are given

Item number											
1	2	3	4	5	6	7	8	9	Hier	k-mean	Two
9	9	9	2	2	2	2	2	2	1	3	1
8	8	6	3	3	3	3	3	3	1	3	1
7	7	7	4	4	4	4	4	4	1	3	1
4	9	9	2	2	6	2	2	2	1	3	1
8	8	8	3	3	3	3	3	3	1	3	1
7	7	7	4	4	4	4	4	4	1	3	1
9	5	9	9	2	2	2	2	2	1	3	1
8	8	8	3	3	3	3	3	3	1	3	1
7	7	7	4	6	4	4	4	4	1	3	1
9	9	9	2	2	2	2	2	2	1	3	1
4	4	4	9	9	9	3	3	3	2	1	3
3	3	3	8	8	8	4	4	4	2	1	3
2	2	2	7	7	7	2	2	2	2	1	3
4	4	4	9	9	9	3	3	3	2	1	3
3	3	3	8	8	8	4	4	4	2	1	3
2	2	2	7	7	7	2	2	2	2	1	3
4	4	4	9	9	9	3	3	3	2	1	3
3	3	3	8	8	8	4	4	4	2	1	3
2	2	2	7	7	7	2	2	2	2	1	3
4	4	4	9	9	9	3	3	3	2	1	3
3	3	3	2	2	2	9	6	9	3	2	2
2	2	2	3	3	3	8	8	8	3	2	2
4	4	4	4	4	4	7	7	7	3	2	2
3	3	3	2	2	2	9	9	2	3	2	2
2	2	2	3	3	3	8	8	8	3	2	2
4	4	4	4	4	4	7	7	7	3	2	2
3	3	3	2	2	2	9	9	9	3	2	2
2	2	2	3	3	3	8	8	8	3	2	2
4	4	4	4	4	4	7	7	7	3	2	2
3	3	3	2	2	2	9	9	9	3	2	2

Figure 9.1 gives a summary of the merging procedure. The average distance estimates are rescaled by the software program to fall into a range of 0–25 units. The patients 21–30 are clustered together at a rescaled distance of approximately 8, while the patients 11–20 and 1–10 are clustered together at respectively 4 and 10 units. When minimizing the output page, the cluster membership is displayed in the original data file. The result for the three cluster solution is given in Table 9.1 (third column from the ride side).

If you wish, you can compute the mean item scores for the separate clusters or draw a graph of the computations. We used SPSS Graphs for that purpose [17].

Table 9.2 The computations of hierarchical clustering are explained underneath. The add-up sum of the squared distances between two patients is first computed for all items. These values are used to estimate the distance between two patients. The smallest distances are used to form the first clusters (see the dendrogram in Fig. 9.1). Then, at larger distances, new clusters are formed by merging the former clusters with additional patients

Item number									Patient
1	2	3	4	5	6	7	8	9	
9.00	9.00	9.00	2.00	2.00	2.00	2.00	2.00	2.00	1
8.00	8.00	6.00	3.00	3.00	3.00	3.00	3.00	3.00	2
Difference									
1.00	1.00	3.00	−1.00	−1.00	−1.00	−1.00	−1.00	−1.00	
Squares									
1.00	1.00	9.00	1.00	1.00	1.00	1.00	1.00	1.00	
Add-up sum of squares									
17.00									

Fig. 9.1 Hierarchical clustering: summary of the merging procedure in the form of dendrogram. The average distance estimates are rescaled by the software program to fall into a range of 0–25 units. The patients 21–30 are clustered together at a rescaled distance of approximately 8, while the patients 11–20 and 1–10 are clustered together at respectively 4 and 10

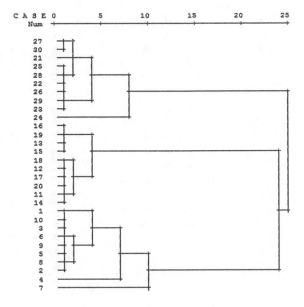

Command: Graphs....Legacy Dialogs....3-D Bar Charts....x-axis: mark Groups of cases....z-axis: mark Separate variables....Define 3-D Bar....Variable: enter cluster membership....Bars represent: enter item scores 1-9....OK.

The graph of Fig. 9.2 gives 27 bars, one for the mean per cluster of each item. It can be observed that in cluster 1 the items 1–3 score highest, in cluster 2 the items 4–6, and in cluster 3 the items 7–9.

We should add, that the cluster identification very well matched the supposed clinical expectation, because the items 1–3 were supposed to score highly in patients

Fig. 9.2 The graphs gives 27 bars, one for each item-mean per cluster. It can be observed that in cluster 1 the items 1–3 score highest, in cluster 2 the items 4–6, and in cluster 3 the items 7–9

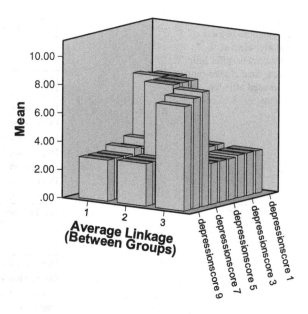

with true depression, the items 4–6 in patients with insomnia, the items 7–9 in patients with reactive depression. And, so, the clusters, as identified, were very well in agreement with the expected clinical diagnoses.

4.2 K-means Clustering

Compared to hierarchical clustering this method works in the opposite direction. It does not start with all patients being a cluster of his/her own, but instead, randomly selects cluster centers, and, then, by iteration, tries and finds the best fit centers for the data given. Intuitively one may assume that this procedure should lead to the same result. However, this is not necessarily true. The point is that one important assumption of the k-means method is that the clusters are equally sized, and this is not an assumption of the hierarchical method. SPSS is used again for analysis.

> Command: Analyze....Classify....K-means Cluster Analysis....Variables: enter item scores 1-9....Label Cases by: patient number as a string variable....Number of clusters: 3 (in our example chosen for comparison with the above method)....click Method: mark Iterate....click Iterate: Maximal Iterations: mark 10....Convergence criterion: mark 0....click Continue....click Save: mark Cluster Membership....click Continue ....click Options: mark Initiate cluster centers....mark ANOVA table ....mark Cluster information for each case....click Continue....OK.

Table 9.3 shows, that the initially randomly-selected cluster-centers differ little from the final cluster-centers as estimated after ten iterations. Table 9.4 shows that the three clusters identified are very significantly different from one another,

Table 9.3 K-means
clustering: the initially
randomly-selected
cluster-centers differ little
from the final cluster-centers
as estimated after ten
iterations

Initial cluster centers

	Cluster		
	1	2	3
Depression score 1	3.00	3.00	9.00
Depression score 2	3.00	3.00	5.00
Depression score 3	3.00	3.00	9.00
Depression score 4	8.00	2.00	9.00
Depression score 5	8.00	2.00	2.00
Depression score 6	8.00	2.00	2.00
Depression score 7	4.00	9.00	2.00
Depression score 8	4.00	9.00	2.00
Depression score 9	4.00	9.00	2.00

Final cluster centers

	Cluster		
	1	2	3
Depression score 1	3.10	3.00	7.60
Depression score 2	3.10	3.00	7.70
Depression score 3	3.10	3.00	7.90
Depression score 4	8.10	2.90	3.60
Depression score 5	8.10	2.90	3.10
Depression score 6	8.10	2.90	3.30
Depression score 7	3.00	8.10	2.90
Depression score 8	3.00	7.80	2.90
Depression score 9	3.00	7.40	2.90

Table 9.4 ANOVA (analysis of variance testing) of the k-means clustering model including three clusters. The three clusters identified are very significantly different from one another, with all of the nine item scores being statistically significant independent predictors of cluster membership

ANOVA

	Cluster		Error			
	Mean Square	df	Mean Square	df	F	Sig.
Depression score 1	69.033	2	1.233	27	55.973	.000
Depression score 2	72.100	2	1.000	27	72.100	.000
Depression score 3	78.433	2	.881	27	88.979	.000
Depression score 4	79.633	2	1.933	27	41.190	.000
Depression score 5	86.800	2	1.063	27	81.659	.000
Depression score 6	83.733	2	1.033	27	81.032	.000
Depression score 7	88.433	2	.733	27	120.591	.000
Depression score 8	78.433	2	.833	27	94.120	.000
Depression score 9	66.033	2	1.900	27	34.754	.000

The F tests should be used only for descriptive purposes because the clusters have been chosen to maximize the differences among cases in different clusters. The observed significance levels are not corrected for this and thus cannot be interpreted as tests of the hypothesis that the cluster means are equal with all of the nine item scores being significant independent predictors of cluster membership. When minimizing the output the data file comes up and the cluster membership is given (Table 9.1 second column from the right). The cluster membership is identical to the one of hierarchical, only the cluster are given a different number. The advantage of k-means is that analysis of variance is provided to test whether the separate item scores are independent contributors to form the separate clusters. In the given example, despite the small size of the data, this was, obviously, true.

4.3   Two-Step Clustering

Two-step clustering is available in SPSS since 2001, and is, particularly, suitable for larger data files as explained above, but it can also be performed with small data as a contrast test for the other methods. It also provides goodness of fit statistics for models with different cluster numbers and a more precise and quantitative assessment of the composition of the clusters and importance of the individual items. This method is called a two-step procedure, because it starts with a large number of preclusters of individuals that are closest to one another. Then, as a second step, a standard hierarchical cluster procedure is followed, not of the individuals, but of the preclusters formed. Finally, the BIC (or AIC, Bayesian or Akaike information criterion, which are, virtually, the same) is used by the software program to enable the investigator to observe at which cluster number the within cluster distances start being adequately small for the investigator's purposes.

> Command: Analyze....Classify....TwoStep Cluster Analysis....Continuous Variables: enter item scores 1-9....Plots....mark Rank of variable importance....Rank Variables: mark By cluster....Importance Measures: mark chi-square or *t*-test....mark confidence intervals....percentage: fill-out 95....Output....Statistics: mark information criterion....Working data file: mark create cluster membership variable....OK.

Table 9.5 gives the autoclustering table of the two-step procedure. It can be observed that with the three-cluster model BIC fell by 55.3, while the ratio of distances was very large, 2.5, and that this was followed by a much smaller value of 1.2. The three-cluster model would, therefore, be an adequate choice for further assessment of the data. If we minimize this output page, we will observe that SPSS has provided again the membership data (Table 9.1, right end column). Again, the result is similar to that of hierarchical clustering, although the order of numbers is again different.

Two-step clustering also examines the mean scores of the clusters. In cluster 1 the items 1–3 score highly, in cluster 3 the items 4–6, and in cluster 2 the items 7–9 do so (Fig. 9.3).

Two-step cluster analysis also assesses the importance of the individual items for the formation of the clusters by calculating the t-values of the individual items per cluster versus a t-value of zero (Fig. 9.4). Normally, a t-value larger or smaller

Table 9.5 The autoclustering table of the two-step procedure. With the three-cluster model BIC fell by 55.3, while the ratio of distances was very large, 2.5, and this was followed by a much smaller value of 1.2. The three-cluster model would, therefore, be an adequate choice for further assessment of the data

Auto-clustering				
Number of clusters	Schwarz's Bayesian Criterion (BIC)	BIC change[a]	Ratio of BIC changes[b]	Ratio of distance measures[c]
1	747.796			
2	731.103	−16.693	1.000	.837
3	727.073	−4.030	.241	2.530
4	782.344	55.271	−3.311	1.236
5	838.984	56.640	−3.393	1.232
6	896.107	57.123	−3.422	1.531
7	952.481	56.374	−3.377	1.071
8	1,007.287	54.806	−3.283	1.105
9	1,062.971	55.684	−3.336	1.949
10	1,122.000	59.030	−3.536	1.080
11	1,180.874	58.874	−3.527	1.015
12	1,241.025	60.151	−3.603	.986
13	1,300.603	59.578	−3.569	2.173
14	1,361.424	60.821	−3.644	1.120
15	1,422.286	60.862	−3.646	[d]

[a]The changes are from the previous number of clusters in the table
[b]The ratios of changes are relative to the change for the two cluster solution
[c]The ratios of distance measures are based on the current number of clusters against the previous number of clusters
[d]The distance at the current number of cluster is zero

than approximately +2 or −2 indicates that the mean result is significantly different from 0, and, thus, that the item is a significant contributor to the formation of the cluster. Here the t-values for rejecting the null hypothesis are taken somewhat larger (approximately 2.7 and −2.7, dotted lines). This is a Bonferroni adjustment for multiple testing. In spite of the adjustment, virtually all of the items remained very significant predictors of being in a cluster.

5   Discussion

In the example of this chapter the three methods produced equally sized clusters, but, with different data, this needs not necessarily be similarly so. Both hierarchical and two-step clustering will allow for clusters of unequal sizes, if a better data fit is provided.

Cluster analysis is, currently, an important methodology in explorative data mining, and a main task in machine learning, but, unfortunately little used in clinical research in spite of the omnipresence of heterogeneities in clinical data files. This

Fig. 9.3 Two-step clustering examines the composition of the clusters by calculating the item score means per cluster (unadjusted 95 % confidence intervals are given). In cluster 1 the items 1–3 score highly, in cluster 3 the items 4–6, and in cluster 2 the items 7–9 do so (only items 1, 4, 7 are shown)

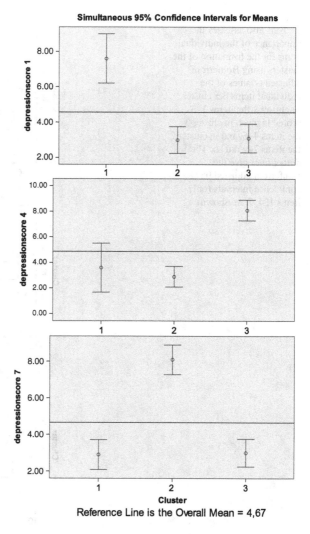

Simultaneous 95% Confidence Intervals for Means

Reference Line is the Overall Mean = 4,67

may be related with the traditional belief in clinical trials in this field of research, where the effect of heterogeneities is minimized through control group observations and need not further taken into account. Controlled clinical trials may, indeed, be more accurate and reliable for making health predictions, but are uncontrolled data completely meaningless? Even if heterogeneities as established are clinically relevant in less than 10 % of the cases, 10 % is better than 0 %. Also 10 % is a lot if you consider the ready availability of large and complex data files in the electronic health records of modern health facilities and other institutions.

The current chapter gives an example of how cluster analysis can be applied in pharmaceutical research. A small data example is given, as this is an explanatory article, but, in practice, very large samples with hundreds of explanatory variables,

Fig. 9.4 Two-step cluster
analysis also assesses the
importance of the individual
items for the formation of the
clusters using Bonferroni
adjusted t-values of the
individual items per cluster.
In cluster 1 the items 1–3
scored highest, in cluster 3
the items 4–6, and in cluster 2
the items 7–9 did so. The
dotted lines give the
Bonferroni adjusted 95 %
confidence intervals (only
items 1, 4, 7 are shown)

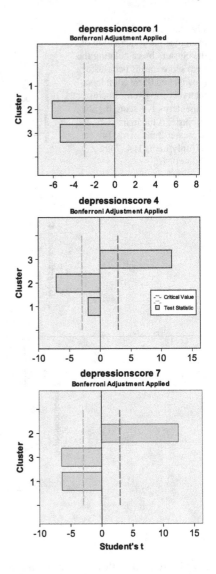

like the over-expressed genes in a micro-array can be included, and can establish
prediction models for future data. A major problem is, that, with large data, the
methods are computationally intensive, and can take weeks or months. Why are,
particularly, hierarchical and k-means clustering so computationally, intensive. This
is, because the clusters are consecutively formed from individuals representing an
individual cluster. These individual clusters including multiple variables are, then,
sequentially merged according to their similarity (= small distance of variable
values). With large datasets computations can take weeks or months even with
modern computers. The time-complexity of a mathematical model is the time

required to run the data. Time is, generally, a so-called polynomial (otherwise called higher order) function of the sample size. According to the big O notation, a term commonly used for the purpose, hierarchical clustering uses O (n^2) polynomial time-complexity, instead of O (n^3), and this is fine, but even so it is slow. An example of an O (n^2) polynomial time equation is the underneath data-time relationship.

$$\text{time} = 3n^2 - 3n + 3$$

as n (data sample size) grows large, the n^2 term will dominate, and so all the other terms can be neglected. For example, if n = 1,000, then the term "$3n^2$" will be 1,000,000 times larger than the "3" term and 1,000 times larger than the "3n" term. Ignoring them, will have a negligible effect on the magnitude of the outcome of the equation. Also, the coefficient of the "$3n^2$" term will become irrelevant, if we compare the equation with a third order equation like n^3. With n = 1,000 the latter term is one billion, the former only three million. The difference between one billion and three million is hardly different from one billion and irrelevant, if you think of a difference expressed in the minutes it takes to complete a data computation. The time-complexity can be minimized by using a two-step cluster analysis, which uses autoclustering statistics, instead of the laborious step up models starting with n individuals and sequentially merging all of the individuals one by one.

We should admit that k-means method is not based on the distances between the individuals, but to the centers of the clusters. If a limited number of clusters is chosen, the computation problem can be somewhat minimized. Otherwise, however, it is similar to that of the hierarchical method. Two-step cluster analysis analyzes the distances between randomly formed small groups of individuals (preclusters), rather than between all of the individuals in the file. A limitation is, however, that a normal distribution is assumed between the numbers of clusters and the overall uncertainty of the model. The model with only one cluster has the largest uncertainty (expressed as standard errors, see Table 9.5), as individuals are very distant from one another. With increasing numbers of clusters the overall standard errors fall. The following uncertainty factor is computed.

Bayesian information criterion (BIC) = n ln standard error^2 + k ln n (n = sample size, ln = natural logarithm, k = numbers of clusters). The largest BIC is observed with the 1 cluster model.

Often the cluster number will be selected, if the BIC rapidly falls and the mean of the distances between the subjects rapidly becomes small. We should add that BIC is a very general goodness of fit test for comparing different mathematical models of any type for analyzing one and the same dataset. It is based on the concept that the standard errors of the models, otherwise called the residuals, follow a normal distribution and that the model with the smallest uncertainty is the best one. It uses, instead of the t-test, the so-called log likelihood test, which has computational advantages, that come about, when taking logs (logarithms) of the equations. But, otherwise, it is similar to the t-test.

We should add that the methods are less adequate to analyze non-Gaussian data and two dimensional (2-D) data like 2-D data in the field of data imaging, for instance, image compression and image color quantization [18–20]. For that purpose another clustering method called density based clustering is more adequate [20].

The clustering methods, as reviewed in this chapter, may be less accurate and reliable than controlled clinical trials, but, on the other hand, they are more flexible, as they do not require a control group or prior randomization of the data. Also, they may, sometimes, be more adequate than multiple regression analysis, as the presence of collinearity (two explanatory variables that are highly correlated with one another) is much less a problem, and the presence of outcome variables is not a requirement.

6   Conclusions

1. Hierarchical, k-means, and two-step cluster analysis can be adequately used for the analysis of heterogeneous clinical data and can provide relevant prognostic patterns.
2. On a theoretical base two-step analysis is better suitable for large data files.
3. The clustering methods as reviewed may be less accurate and reliable than controlled clinical trials, but, on the other hand, they are more flexible, because they do not require a control group or prior randomization of the data.
4. Also, they may, sometimes, be more adequate than multiple regression analysis, because the presence of collinearity (two explanatory variables that are highly correlated with one another) is much less a problem, and the presence of an outcome variable is not required.

References

1. Anonymous (1978) Diagnostic and statistical manual of mental disorders. American Psychiatric Society, New York
2. Solyanik GI (2010) Multifactorial nature of tumor drug resistance. Exp Oncol 32:181–185
3. National Institutes of Health, 9000 Rockville Pike, Bethesda, Maryland (2003) Enhancing HIV vaccine efficacy in high-risk drug users. Release data Jan 6, RFA Number DA-03-002
4. Alsford S, Eckert S, Baker N, Glover L, Sanchez-Flores A, Leung KF, Turne DJ, Field MC, Berriman M, Horn D (2012) High throughput decoding of antitrypanosomal drug efficacy and resistance. Nature. doi:10.1038/nature 10771
5. Anonymous (2012) Machine learning. http://en.wikipedia.org/Machine_learning. 12 July 2012
6. Anonymous (2012) Cluster analysis. http://en.wikipedia.org/Cluster_analysis. 12 July 2012
7. Anonymous (2012) Sequence clustering. http://en.wikipedia.org/wiki/Sequence_analysis. 12 July 2012
8. Kim HK, Choi IJ, Kim HS, Oshima A, Michalowski A, Green JE (2011) A gene expression signature of acquired chemoresistance to cisplatinum and fluorouracil combination chemotherapy in gastric cancer patients. PLoS One 18:e16694

9. Yeh ST (2012) Clinical adverse event data analysis and visualization. Smith Kline Datafile, 10 July 2012

10. Bychowiec B, Piskorski J, Stanislawska K, Dziarmaga M, Mineczykowski A, Wykretowicz A, Wysocki H (2010) An exploratory clustering study of rare adverse events in drug deluting stent patients. Comput Methods Sci Technol 16:5–11

11. Xu D, Redamn-Furey N (2007) Statistical cluster analysis of pharmaceutical solvents. Int J Pharm 339:175–188

12. Hawwa A, Millership JS, Collier PS, McCarthy A, Dempsey S, Cairns C, McElnay JC (2009) Development of objective methodology to measure medication adherence to oral thiopurines in paediatric patients with acute lymphoblastic leucemia. Eur J Clin Pharmacol 65:1105–1112

13. Sibson R (1973) An optimally efficient algorithm for a single-link cluster method. Comput J 16:30–34

14. Defays D (1977) An efficient algorithm for a complete link cluster method. Comput J 20:364–366

15. Lloyd SP (1982) Least square quantization in PCM. IEEE Trans Inf Theory 28:129–137

16. Banfield JD, Raftery AE (1993) Model-based Gaussian and non-Gaussian clustering. Biometrics 49:803–821

17. SPSS statistical software (2012) www.spss.com. 3 Oct 2012

18. Ester M, Kriegel HP, Sander J, Xu X (1996) A density based algorithm for discovering cluster in large spatial databases with noise. In: Proceedings of 2nd international conference on knowledge discovery and data mining. AAI Press, Portland

19. Data Clustering Applets (2012) http://webdocs.cs.ualberts.ca/~yaling/Cluster/applet. 17 Sept 2012

20. Anonymous (2012) Density-based cluster and outlier analysis. www.dbs.informatik.uni-muenchen.de/Forschung/KDD/Clustering. 25 Sept 2012

Chapter 10
Anomaly Detection

1 Summary

1.1 Background

With large data files outlier recognition requires a more sophisticated approach than the traditional data plots and regression lines. Also, the number of outliers tends to rise linearly with the data's sample size.

1.2 Objective

To examine whether BIRCH (balanced iterative reducing and clustering using hierarchies) clustering is able to detect previously unrecognized outlier data.

1.3 Methods

A simulated and a real data file were used as examples. SPSS statistical software was used for data analysis.

1.4 Results

1. In 50 mentally depressed persons a regression analysis failed to detect any outliers. BIRCH analysis of these data showed in addition to two clusters a relevant outlier cluster consistent of seven patients (14 %) not fitting in the formed clusters.

T.J. Cleophas and A.H. Zwinderman, *Machine Learning in Medicine:*
Part Two, DOI 10.1007/978-94-007-6886-4_10,
© Springer Science+Business Media Dordrecht 2013

2. In 576 iatrogenic admissions the number of co-medications was not a significant loglinear predictor of the iatrogenic admission. In contrast, BIRCH analysis revealed an outlier cluster consistent of 174 patients (30 %) with extremely many co-medications.

1.5 Conclusions

1. A systematic assessment for outliers is important in therapeutic research with large data, because the lack of it can lead to catastrophic consequences.
2. Traditional data analysis, like regression analysis, was unable to demonstrate outliers in our examples.
3. BIRCH cluster analysis of our examples produced relevant outlier clusters of patients not fitting in the data otherwise.
4. On theoretical grounds BIRCH cluster analysis is, particularly, suitable for large datasets.

2 Introduction

Graphs like data plots and regression lines are convenient for visualizing outliers in therapeutic data patterns, and have been successfully used for that purpose for centuries. They are, however, arbitrary, and, with large data files, both data pattern and outlier recognition require a more sophisticated approach. Also, the number of outliers, generally, tends to rise linearly with the sample size. BIRCH is the abbreviation of "balanced iterative reducing and clustering using hierarchies" [1]. It was introduced by three computer scientists from Wisconsin University [1] in 1996, and is available in SPSS's module Classify, under "two-step cluster analysis", since 2001. It is an unsupervised data mining methodology suitable for very large datasets, but can also be applied for small data [2, 3]. It is, currently, mainly used by econo- and sociometrists [3], and, like other machine learning methods, little used in therapeutic research. This is, probably, due to the traditional belief of clinicians in clinical trials where outliers are assumed to be equally balanced by the randomization process and are not further taken into account. In contrast, modern computer data files often involve large uncontrolled data files, and arbitrary methods like scatter plots do not adequately detect outliers in the data.

The current chapter, using a simulated and a real data example, examines whether BIRCH clustering is able to detect previously unrecognized outlier data. Step by step analyses were performed for the convenience of investigators. This chapter was also written as a hand-hold presentation accessible to clinicians and a must read publication for those new to the method.

3 Identifying Outliers Using BIRCH Clustering, Some Theory

We will use two-dimensional data for simplicity, but multidimensional data can be applied as well. With traditional clustering methods like hierarchical [2] and k-means clustering [3] clusters are identified by computing their data distances, taking their differences along the x and y-axes. BIRCH clustering uses sums of squares of the x- and y-values to summarize the data of the clusters. It also uses repeated binary partitions of the data to form clusters with each cluster having its own metrics, in terms of size and sum of squares. By iteration the software tries and finds the best fit metrics for the data given, meaning, that clusters can be split and thresholds for forming new clusters are chosen. The computation is rather complex, because all possible combinations of the data are checked by the computer. For example, with only four data six clusters of two are possible, with 100 data 5,000 are possible, etc. Large clusters, generally, produce large metrics, small clusters produce small metrics, and it not only depends on the size of the metrics, but also on the investigators' preferences, which numbers of clusters will ultimately be chosen for further data interpretation. The split clusters can be viewed as the branches of a tree and binary partitioning of the branches increases the height of the tree. The higher the tree, the more time the clustering operation takes.

A major problem of clustering analysis is that it is time-costly and may run out of computer memory. Sometimes, memory allocation, using additional computers, is the only solution. As an example, a tree with binary branches can contain $2^0 + 2^1 + 2^2 + 2 \ldots = 2^{h+1} - 1$ branches on top of one another, where h = the number of branch layers. If we neglect the "1" terms, we will find $h > {}^2\log(n)$. With n = number of branches = 1,000, a tree of ten branch layers would be mostly cost-efficient in terms of computing time/required computer memory. BIRCH clustering manages to keep the height of the tree small, and uses for that purpose cluster rotation by moving branches with multiple clusters up and those with few clusters down.

For the identification of outliers BIRCH applies a-priori given tree capacities. If the tree is full, and cannot further accept any patients, then the patients with the worst fit to the formed clusters will be moved into a so-called noise cluster, otherwise called outlier cluster.

SPSS also offers advanced options: tree capacities can be somewhat improved by changing metrics. This should, however, be handled with care, as it can easily lead to loose and meaningless clusters, and loss of system performance.

4 Simulated Data Example

A study of 50 mentally depressed patients is used. Age and depression severity scores (1 for mild and 10 for severest depression) are given in the first and second column (Table 10.1). Linear regression between the two variables gave

Table 10.1 A study of 50 mentally depressed patients. Age and depression severity scores (1 for mild and 10 for severest depression) are given in the first and second column. The cluster membership computed by two step BIRCH clustering is in column 3: two clusters were identified (indicated with 1 and 2) and one outlier cluster (indicated with −1)

Age	Depression score	Cluster membership
20.00	8.00	2
21.00	7.00	2
23.00	9.00	2
24.00	10.00	2
25.00	8.00	2
26.00	9.00	2
27.00	7.00	2
28.00	8.00	2
24.00	9.00	2
32.00	9.00	2
30.00	1.00	−1
40.00	2.00	−1
50.00	3.00	1
60.00	1.00	−1
70.00	2.00	1
76.00	3.00	1
65.00	2.00	1
54.00	3.00	1
54.00	4.00	1
49.00	3.00	1
30.00	4.00	2
25.00	5.00	2
24.00	4.00	2
27.00	5.00	2
35.00	6.00	2
45.00	5.00	1
45.00	6.00	2
67.00	7.00	1
80.00	6.00	1
80.00	5.00	1
40.00	1.00	−1
50.00	2.00	1
60.00	3.00	1
80.00	4.00	1
50.00	5.00	1
76.00	6.00	1
65.00	7.00	1
79.00	8.00	−1
57.00	3.00	1
46.00	4.00	1
54.00	5.00	1
74.00	6.00	1
65.00	7.00	1
57.00	9.00	−1

(continued)

Table 10.1 (continued)

Age	Depression score	Cluster membership
68.00	8.00	−1
67.00	7.00	1
65.00	6.00	1
64.00	5.00	1
74.00	4.00	1
75.00	3.00	1

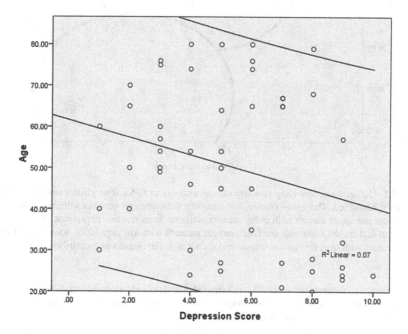

Fig. 10.1 Graph of the data from the example with linear regression line and the 90 % confidence intervals: the 90 % confidence intervals produced no more than a single case very close to the intervals boundary, but otherwise no hint of any outliers

some evidence for a weak negative correlation between the two with p = 0.063. This would be compatible with the concept that younger are more at risk of high severity due to true depression, the older are so of low severity due to reactive depression. However, in case-reviews outlier forms of depression like insomnia groups have been noted, but no hints of such is given in the regression model. Even the 90 % confidence intervals produced no more than a single case very close to the intervals boundary (Fig. 10.1). An outlier analysis using two step BIRCH analysis was performed. SPSS statistical software was used for analysis [3].

Command: Analyze....Classify....Two Step Cluster Analysis ....Continuous Variables: enter age and depression score....Distance Measure: mark Euclidean....Clustering Criterion: mark Schwarz's Bayesian Criterion....click Options: mark Use noise

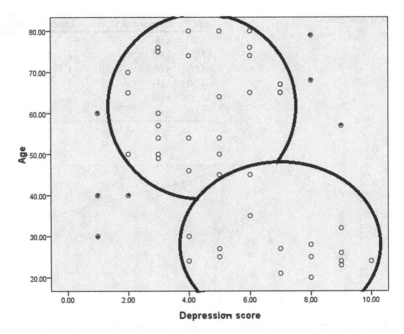

Fig. 10.2 Outlier detection using two step cluster analysis in SPSS. Two cluster and one outlier data set is identified. The *lower cluster* is compatible with younger patients suffering from true depression, the *upper cluster* with older patients suffering from reactive depression. The outliers on the *left* and on the *right side* are four younger patients with low depression scores, and three older patients with high depression scores, and do not fit in the clusters as established

handling ....percentage: enter 25....Assumed Standardized: enter age and depression score....click Continue....click Output: Working Data File: mark Create cluster membership variable....click Continue....click OK.

When returning to the data file, it now shows the cluster membership of each case 1–50 (third column). Two clusters have been identified (indicated by 1 and 2) and one outlier cluster (indicated by −1). We use SPSS again to draw a dotter graph of these results.

Command: Analyze....Graphs....Legacy Dialogs: click Simple Scatter ....Define.... Y-axis: enter Age....X-axis: enter Depression score....OK.

Figure 10.2 shows two clusters with oval and, because of the similarly sized scales, even approximately round patterns. They are also approximately similar in size but this needs not to be so. Also, seven outlier data are shown. The results do very well match the patterns as clinically expected: two populations, one with younger and severely patients with true depression and one with older and milder depressed patients with only a reactive depression. The outliers consist of seven patients of all ages not fitting in the formed clusters. They may suffer from insomnia or other rare forms of the depression syndrome.

5 Real Data Example

In a 2,000 patient study of hospital admissions 576 possibly iatrogenic were identified by a team of specialists [4]. The number of concomitant medications (co-medications) was not a significant predictor of hospital admission in the logistic regression of the data, but when transformed into a categorical factor it was [5]. In order to find an explanation for this finding, a BIRCH two step cluster analysis of these data was performed in SPSS [3].

> Command: Analyze....Classify....Two Step Cluster Analysis ....Continuous Variables: enter age and co-medications....Distance Measure: mark Euclidean....Clustering Criterion: mark Schwarz's Bayesian Criterion....click Options: mark Use noise handling ....percentage: enter 25....Assumed Standardized: enter age and co-medications....click Continue....click Plot: mark Cluster pie chart....click Continue....click Output: Statistics....mark Descriptives by cluster....mark Cluster frequencies....mark Information Criterion ....Working Data File: mark Create cluster membership variable....click Continue....click OK.

Table 10.2 gives the autoclustering table of the two-step BIRCH procedure. It can be observed that 15 different models are assessed (including 1–15 clusters). This table shows something about the precision of the different models, as estimated by the overall uncertainties (or standard errors) of the models (measured by Schwarz's Bayesian Criterion (BIC) = n ln (standard error)^2 + k ln n, where n = sample size, ln = natural logarithm, k = number of clusters). With the three or four cluster models the smallest BIC was observed, and, thus, the mostly precise model. The three or four cluster model, including an outlier cluster, would, therefore, be an adequate choice for further assessment of the data. The Tables 10.3 and 10.4 gives description and frequency information of the four cluster model. Figure 10.3 gives a pie chart of the size of the three clusters and the outlier cluster. If we minimize this output page, and return to the data file, we will observe that SPSS has provided again the membership data. This file is too large to understand what is going on, and, therefore we will draw a three dimensional graph of this output.

> Command: Graphs....Legacy Dialogs....3 D Bar Charts....X-axis represents: click Groups of cases....Y- axis represents: click Groups of cases....click Define....Variable: enter co-medications....Bars represent: enter mean of values....X-Category axis: enter age....Y-Category axis: enter two step cluster number variable....click OK.

Figure 10.4 shows the result. In front two clusters with younger patients and few co-medications are observed. In the third row is one cluster of elderly with considerably more co-medications. Then, at the back the patients are who do not fit in any of the clusters. They are of all ages, but their numbers of co-medications are generally very high. This finding is relevant, because it supports a deleterious effect of numbers of co-medications on the risk of iatrogenic admission.

Table 10.2 Fifteen different cluster models have been assessed by the two-step BIRCH procedure (including 1–15 clusters). The precision of the different models, as estimated by the overall uncertainties measured by Schwarz's Bayesian Criterion (BIC) is given. With the three or four cluster models the smallest BIC was observed, and, thus, the mostly precise model

Auto-clustering				
Number of clusters	Schwarz's Bayesian Criterion (BIC)	BIC change[a]	Ratio of BIC changes[b]	Ratio of distance measures[c]
1	293.899			
2	277.319	−16.580	1.000	1.513
3	185.362	−91.957	5.546	1.463
4	178.291	−7.071	.426	1.007
5	197.946	19.654	−1.185	1.141
6	216.403	18.457	−1.113	1.159
7	236.467	20.064	−1.210	1.099
8	251.072	14.606	−.881	1.629
9	272.582	21.509	−1.297	1.125
10	291.641	19.059	−1.150	1.015
11	301.090	9.449	−.570	1.000
12	308.019	6.929	−.418	1.058
13	321.943	13.924	−.840	1.197
14	339.382	17.439	−1.052	1.074
15	361.262	21.880	−1.320	1.225

[a]The changes are from the previous number of clusters in the table
[b]The ratios of changes are relative to the change for the two cluster solution
[c]The ratios of distance measures are based on the current number of clusters against the previous number of clusters

Table 10.3 Description information of the four cluster model selected from the 15 models from Table 10.2

Centroids		Age		Comed	
		Mean	Std. deviation	Mean	Std. deviation
Cluster	1	1,928.9227	6.50936	2.5028	.50138
	2	1,933.7171	6.01699	.6250	.48572
	3	1,956.8551	6.16984	1.0725	.64895
	Outlier (−1)	1,939.7644	20.15623	2.4138	1.75395
	Combined	1,936.8090	14.91570	1.8090	1.34681

6 Discussion

There is no rigorous mathematical definition for outliers of a dataset, unlike there is for, for example, p-values, r-values etc. Why then worry about the outliers after all? This is, because they can lead not only to serious misinterpretations of the data, but also to catastrophic consequences once the data are used for making predictions, like serious and, sometimes, even fatal adverse events from drug treatments.

Table 10.4 Frequency information of the four cluster model selected from the 15 models from Table 10.2

Cluster distribution		N	% of combined	% of total
Cluster	1	181	31.4	9.1
	2	152	26.4	7.6
	3	69	12.0	3.5
	Outlier (−1)	174	30.2	8.7
	Combined	576	100.0	28.8
Excluded cases		1,424		71.2
Total		2,000		100.0

Fig. 10.3 Pie chart of the cluster size of the four cluster model selected from the 15 models from Table 10.2

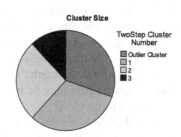

Fig. 10.4 Three-dimensional bar chart of the four cluster model selected from the 15 models from Table 10.2. Over 100 bars indicating mean numbers of co-medications in age classes of 1 year. In the clusters 2 and 3 the patients are young and have few co-medications, in the cluster 1 the patients are old and have many co-medications, in the outlier cluster all ages are present and exceptionally high numbers of co-medications are frequently observed

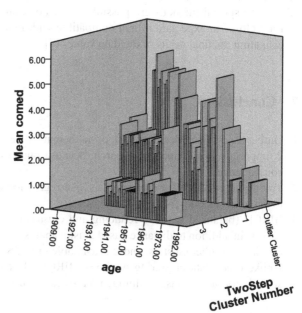

The current chapter shows that traditional methods like regression analysis is often unable to demonstrate outliers, while outlier detection using BIRCH two step clustering is more successful to that aim. We should add that this clustering method points to remote points in the data and flags them as potential outliers. It does not confirm any other prior expectation about the nature or pattern of the outliers. The outliers, generally, involve both extremely high and extremely low values. The approach is, obviously, explorative, but, as shown in the examples, it can produce interesting findings, and theories, although waiting for confirmation. Other forms of cluster analysis include hierarchical, k-means and density-based clustering [2]. Although they can produce multiple clusters, they do not explicitly allow for an outlier option. Nonetheless, investigators are, of course, free to make interpretations about outlier clusters from the patterns as presented.

This chapter only addresses two-dimensional data (one x and one y-variable), but, similarly to multiple regression, BIRCH analysis can be used for analyzing multi-dimensional data, although the computations will rapidly become even more laborious and computer memory may rapidly fall short. In the future this kind of research will be increasingly performed through a network of computer systems rather than a single computer system let alone standalone computers. Also, multi-dimensional outliers may be harder to interpret, because they are associated with multiple factors.

This chapter addresses only outlier-assessment in data without outcome variables. If outcome variables are available, other methods can be used, particularly, the identification of data beyond the confidence limits of the outcome variables. Also some special methods are possible, then. For example, looking for the data that are closer to expectation than compatible with random distributions [6], and investigating the final digits of the data values [7].

7   Conclusions

1. Outlier assessment, despite a lack of mathematical definition of the term outliers, is important in therapeutic research, because outliers can lead to catastrophic consequences.
2. Traditional methods for data analysis, like regression analyses, are often unable to demonstrate outliers.
3. BIRCH (balanced iterative reducing and clustering using hierarchies) analysis allows, in addition to distance based cluster analysis, for the identification option of relevant outlier clusters, and it is available in SPSS statistical software.
4. Unlike other data clustering methods, BIRCH cluster analysis is, particularly, suitable for large datasets, although it also can be used for small data.

References

1. Zhang T, Ramakrishnan R, Livny M (1996) BIRCH. In: SIGMOD anthology (The organ of the "special interest group on management of data" of the Association of Computing Machinery), vol 6, pp 102–114
2. Mooi E, Sarstedt M (2011) Cluster analysis. In: A concise guide to market research. Springer, Berlin/Heidelberg, pp 237–284
3. SPSS Statistical Software (2012) www.spss.com. 8 Oct 2012
4. Atiqi R, Cleophas TJ, Van Bommel E, Zwinderman AH (2010) Prevalence of iatrogenic admissions to the departments of medicine/cardiology/pulmonology in a 1250 beds general hospital. Int J Clin Pharmacol Ther 48:517–521
5. Cleophas TJ, Atiqi R, Zwinderman AH (2011) Handling categories properly: a novel objective of clinical research. Am J Ther. doi:10.1097
6. Cleophas TJ, Zwinderman AH (2012) Testing clinical trials for randomness. In: Statistics applied to clinical studies. Springer, Dordrecht, pp 469–479
7. Cleophas TJ, Zwinderman AH (2012) Research closer to expectation than compatible with random sampling. In: Statistics applied to clinical studies. Springer, Dordrecht, pp 139–147

References

Chapter 11
Association Rule Analysis

1 Summary

1.1 Background

Regression analysis and other traditional methods for assessing risk prediction are not sensitive with weak associations.

1.2 Objective

To assess whether association rule analysis performs better for the purpose.

1.3 Methods

Hypothesized data samples were used as examples.

1.4 Results

The fraction of patients with both coronary artery disease (CAD) and overweight was significantly larger in the patients with CAD than it was in the entire population, rising from 0.2 to 0.5, $p < 0.001$. The observed proportion of patients with both overweight and CAD compared to the expected proportion if overweight and CAD would have occurred independently of one another, was larger than expected, 1.25 versus 1.00 ($p < 0.001$). Similar results were observed when two predictors instead of one were used. The traditional tests included binary logistic regression, McNemar's and Cochran's tests. All of them were statistically insignificant.

T.J. Cleophas and A.H. Zwinderman, *Machine Learning in Medicine:*
Part Two, DOI 10.1007/978-94-007-6886-4_11,
© Springer Science+Business Media Dordrecht 2013

1.5   Conclusions

1. Association rule analysis was more sensitive than regression analyses and paired chi-square tests, and was able to demonstrate significant predictor effects, when the other methods did not.
2. It can include multiple variables and very large datasets. It seems to be a welcome novel methodology for clinical predictor research.
3. It only uses part of the data, and some of the information may, therefore, be lost, and, as an exploratory method of research, it is at increased risk of false positive results and unexpected results.

2   Introduction

In clinical research the effect of a predictor on health/disease is often the study's subject. Examples include the effect of vaccination on influenza and the effect of overweight on coronary artery disease (CAD). Traditionally, regression analyses with the assumed predictor as independent variable, paired chi-square tests, and other methods are applied for significance testing of the assumed effects. However, these methods are only sensitive with strong associations, and, with weak associations, association rule analysis based on conditional probabilities may be more sensitive, because the latter uses subsets of the data rather than the entire dataset. Based on the concept of conditional probabilities, Rakesh Agrawal, technical fellow of Microsoft Research from IBM Almaden Research Center, San Jose, California, introduced in 1993 association rules for discovering associations between the sales data of supermarkets, like the sale of onions and potatoes in relation with the sale of burgers [1]. Currently, mining association rule methodology is an important task of machine learning, a novel discipline which replaces mental arithmetic with artificial intelligence and uses specialized statistical software for the purpose. Mining association rules are already widely used in marketing research [2], and in econo-[3] and sociometry [4]. However, it is little used in clinical research, despite the need for better health prediction in this field, particularly in cardiovascular research [5–7]. When searching Medline, we found only one bibliographic study reviewing Medline publications [8], one review study of traditional Chinese medicine [9], and one theoretical study of association rules for identifying relevant patterns in medical databases [10], but no real data studies of the latter. The current paper using simulated data assesses the sensitivity of the Agrawal association rules as compared with the traditional methods for testing, and uses simple confidence intervals for testing significance [11]. Although (free) software is available for large datasets [12–14], small datasets can be analyzed using a pocket calculator. This will be demonstrated. This chapter was written as a hand-hold presentation accessible to clinicians and a must-read publication for those new to the method.

3 Some Theory

Association rule learning is based on the following approach. A representative data sample, otherwise called learning sample, is taken to identify rules for making predictions about future patients. A small example is used for illustration purposes (1 = yes, 0 = no). Proportions observed are equal to chances or probabilities. If you observe a 40 % proportion of healthy patients, then the chance or probability (P) of being healthy in this group is 40 %. With two variables, e.g., healthy and happy, the symbol ∩ is often used to indicate "and" (both are present). Underneath a hypothesized example of five patients, with three of them having overweight and two of them coronary artery disease (CAD), is given.

	Overweight (predictor)	Coronary artery disease
Patient	X	Y
1	1	0
2	0	1
3	0	0
4	1	1
5	1	0

Support rule

$$\text{Support} = P\, X \cap Y = 1/5 = 0.2$$

Confidence rule

$$\text{Confidence} = P\, X \cap Y\, /\, P\, Y = [1/5]\, /\, [2/5] = 0.5$$

Lift rule (or lift-up rule)

$$\text{Lift} = P\, X \cap Y\, /\, [P\, X \times P\, Y] = [1/5]\, /\, [2/5 \times 2/5] = 1.25$$

Conviction rule

$$\text{Conviction} = [1 - P\, Y]\, /\, [1 - P\, X \cap Y\, /\, P\, Y] = [1 - 2/5]\, /\, [1 - 0.5] = 1.20$$

The rules may look complex, but their interpretation is pretty straightforward.

I. The support gives the proportion of patients with both overweight and CAD in the entire population.
II. The confidence gives the fraction of patients with both CAD and overweight in those with CAD. This fraction is obviously larger than that in the entire population, because it rose from 0.2 to 0.5.

III. The lift compares the observed proportion of patients with both overweight and CAD with the expected proportion if CAD and overweight would have occurred independently of one another. Obviously, the observed value is larger than expected, 1.25 versus 1.00, suggesting that overweight does contribute to the risk of CAD.

IV. Finally, the conviction compares the patients with no-CAD in the entire population with those with both no-CAD and the presence of overweight. The ratio is larger than 1.00, namely 1.20. Obviously, the benefit of no-CAD is better for the entire population than it is for the subgroup with overweight.

In order to assess whether the computed values, like 0.2 and 1.25, are significantly different from 0.0 to 1.0 confidence intervals have to be calculated. We will use the McCallum-Layton calculator for proportions, freely available from the Internet [11]. The calculations will somewhat overestimate the true confidence intervals, because the true confidence intervals are here mostly composed of two or more proportions, and this is not taken into account. Therefore, the p-values are somewhat larger than they should be. However, if significant, we need not worry.

4   Examples

Table 11.1 gives the first example. 20/50 of the patients have overweight (predictor), 20/50 have CAD. A paired binary test (McNemar's test) shows no significant difference between the two columns ($p = 1.0$). Binary logistic regression with the predictor as independent variable is equally insignificant ($b = 0.69, p = 0.241$).

Applying the Agrawal association rules we find a support of 0.2 and confidence of 0.5. The lift is 1.25 and the conviction is 1.20. The McCallum calculator gives the confidence intervals, respectively 10–34 %, 36–64 %, 110–145 %, and 107–137 %. All of these 95 % confidence intervals indicate a very significant difference from respectively 0 % (support and confidence) and 100 % (lift and conviction) with p-values < 0.001. Indeed, the predictor overweight had a very significant positive effect on the risk of CAD.

Table 11.2 gives the second example; the risk of both overweight and being a manager on the risk of coronary artery disease is assessed. Instead of a single x –variable now two of them are included. 30/60 of the patients have overweight, 40/60 are manager, and 30/60 have CAD. A paired binary test (Cochran's test) shows no significant difference between the three columns ($p = 0.082$). Binary logistic regression with the two predictors as independent variables is equally insignificant (b-values are − 21.9 and 21.2, p-values are 0.99 and 0.99).

Applying the Agrawal association rules we find a support of 0.1666 and confidence of 0.333. The lift is 2.0 and the conviction is 1.25. The McCallum calculator gives the confidence intervals. Expressed as percentages they are respectively, 8–29 %, 22–47 %, and 159–270 % and 108–136 %. All of these 95 % confidence

Table 11.1 A data set of patients with coronary artery disease or not (1 = yes) and overweight as predictor (1 = yes)

Patient	Overweight	Coronary artery disease
1	1.00	0.00
2	0.00	1.00
3	0.00	0.00
4	1.00	1.00
5	0.00	0.00
6	1.00	0.00
7	0.00	1.00
8	0.00	0.00
9	1.00	1.00
10	0.00	0.00
11	1.00	0.00
12	0.00	1.00
13	0.00	0.00
14	1.00	1.00
15	0.00	0.00
16	1.00	0.00
17	0.00	1.00
18	0.00	0.00
19	1.00	1.00
20	0.00	0.00
21	1.00	0.00
22	0.00	1.00
23	0.00	0.00
24	1.00	1.00
25	0.00	0.00
26	1.00	0.00
27	0.00	1.00
28	0.00	0.00
29	1.00	1.00
30	0.00	0.00
31	1.00	0.00
32	0.00	1.00
33	0.00	0.00
34	1.00	1.00
35	0.00	0.00
36	1.00	0.00
37	0.00	1.00
38	0.00	0.00
39	1.00	1.00
40	0.00	0.00
41	1.00	0.00
42	0.00	1.00
43	0.00	0.00

(continued)

Table 11.1 (continued)

Patient	Overweight	Coronary artery disease
44	1.00	1.00
45	0.00	0.00
46	1.00	0.00
47	0.00	1.00
48	0.00	0.00
49	1.00	1.00
50	0.00	0.00

Table 11.2 A data set of patients with coronary artery disease or not (1 = yes) with overweight and "being manager" as predictors (1 = yes)

Patient	Overweight	Manager	Coronary artery disease
1	1.00	1.00	0.00
2	0.00	0.00	1.00
3	1.00	1.00	0.00
4	0.00	1.00	1.00
5	0.00	0.00	0.00
6	1.00	1.00	1.00
7	1.00	1.00	0.00
8	0.00	0.00	1.00
9	1.00	1.00	0.00
10	0.00	1.00	1.00
11	0.00	0.00	0.00
12	1.00	1.00	1.00
13	1.00	1.00	0.00
14	0.00	0.00	1.00
15	1.00	1.00	0.00
16	0.00	1.00	1.00
17	0.00	0.00	0.00
18	1.00	1.00	1.00
19	1.00	1.00	0.00
20	0.00	0.00	1.00
21	1.00	1.00	0.00
22	0.00	1.00	1.00
23	0.00	0.00	0.00
24	1.00	1.00	1.00
25	1.00	1.00	0.00
26	0.00	0.00	1.00
27	1.00	1.00	0.00
28	0.00	1.00	1.00
29	0.00	0.00	0.00
30	1.00	1.00	1.00
31	1.00	1.00	0.00
32	0.00	0.00	1.00
33	1.00	1.00	0.00
34	0.00	1.00	1.00

(continued)

Table 11.2 (continued)

Patient	Overweight	Manager	Coronary artery disease
35	0.00	0.00	0.00
36	1.00	1.00	1.00
37	1.00	1.00	0.00
38	0.00	0.00	1.00
39	1.00	1.00	0.00
40	0.00	1.00	1.00
41	0.00	0.00	0.00
42	1.00	1.00	1.00
43	1.00	1.00	0.00
44	0.00	0.00	1.00
45	1.00	1.00	0.00
46	0.00	1.00	1.00
47	0.00	0.00	0.00
48	1.00	1.00	1.00
49	1.00	1.00	0.00
50	0.00	0.00	1.00
51	1.00	1.00	0.00
52	0.00	1.00	1.00
53	0.00	0.00	0.00
54	1.00	1.00	1.00
55	1.00	1.00	0.00
56	0.00	0.00	1.00
57	1.00	1.00	0.00
58	0.00	1.00	1.00
59	0.00	0.00	0.00
60	1.00	1.00	1.00

intervals indicate a very significant difference from respectively 0 % (support and confidence) and 100 % (lift and conviction) with p-values < 0.001. Indeed, the predictors overweight and being manager had a statistically very significant effect on the risk of CAD.

5 Discussion

Traditionally, regression analyses with the assumed predictor as independent variable or paired chi-square tests are applied for significance testing of the assumed effects of predictors on health/disease. Association rule analyses give conditional probabilities and were much more sensitive than the traditional methods for demonstrating such assumed effects.

The current paper uses one and two predictor examples only, but the methods can be applied with multiple predictor data. However, computations are more complex and statistical software is required. The software for such purposes are widely available, and free software is in the references [12–14].

Some limitations of association rule mining have to be mentioned. First, it may be more sensitive, but it uses subsets of the data rather than the entire dataset. E.g., the confidence rule and the lift rule are based on only 40 % of the data. Second, like regression analysis association rule analysis is explorative rather than confirmative research, and with multiple variables, a risk of false positive results, otherwise called type 1 errors, is increased. Third, like with other exploratory methods such as regression analysis there is the risk of spurious associations, particularly if associations are unexpected. A classical example is the association of beer and baby diaper sales in a supermarket chain. How can drinking beer and buying diapers be associated? The authors have tried to explain the unexpected finding, but the possibility of a spurious association or type 1 error could not be ruled out [15].

We should add that four rules have been reported in this chapter, but there is more. Additional association rule methods include the all-confidence rule, collective strength rule, leverage rule, and several more [16].

6   Conclusions

1. Association rule analysis is more sensitive than regression analysis and paired chi-square tests, and is able to demonstrate significant predictor effects, when the other methods are not.
2. It can include multiple variables and very large datasets, and statistical software is widely available. It seems a welcome methodology for clinical predictor research.
3. It only uses part of the data, and information may, therefore, easily be lost.
4. As en exploratory method of research, it is at increased risk of false positive results and unexpected results.

References

1. Agrawal R, Imielinski T, Swami A (1993) Mining association rules between sets of items in large data bases. In: SIGMOD anthology (The organ of the "special interest group on management of data" of the Association of Computing Machinery), vol 3, pp 207–216
2. Webb G (2007) Discovering significant patterns. In: Machine learning. Springer, Dordrecht, pp 1–33
3. Kavitha K, Ramaraj E (2012) Mining actionable patterns using combined association rules. Int J Curr Res 4:117–120
4. Bonchi F, Castillo C, Gionis A, Kaimes A (2011) Social network analysis and mining for business applications. ACM Trans Intell Syst Technol 22:1–37
5. Gender TS, Steyerberg EW, Hunnink MG (2012) Prediction model to estimate presence of coronary artery disease: retrospective pooled analysis of existing cohorts. BMJ. doi:10.1136/bmj.e3485
6. Sivasankaran S, Nair MK, Babu G, Zufikar AM (2011) Need for better anthropometric markers for prediction of cardiovascular risk in nutritionally stunted populations. Indian J Med Res 133:557–559

7. Stern MP, Fatchi P, Williams K, Haffner SM (2002) Predicting future cardiovascular disease. Do we need the oral glucose tolerance test? Diabetes Care 25:1851–1856

8. Hristovski D, Stare J, Pterlin B, Dzeroski S (2001) Supporting discovery in medicine by association rule mining in Medline and UMLS. In: Patel VL et al (eds) MEDINFO. IOS Press, Amsterdam, pp 1344–1354

9. Feng Y, Wu Z, Zhou X, Zhou Z, Fan W (2006) Methodological reviews: knowledge discovery in traditional Chinese medicine: state of the art and perspectives. J Artif Intell Med 38:219–236

10. Stilou S, Bamidis PD, Maglaveras N, Pappas C (2001) Mining association rules from clinical databases: an intelligent diagnostic process in healthcare. In: Patel VL et al (eds) MEDINFO. IOS Press, Amsterdam, pp 1399–1409

11. Confidence interval calculator for proportions (2012) www.mccallum-layton.co.uk/. 20 Oct 2012

12. SIPINA free software for academic data mining (2012) http://eric.univ-lyon2.fr/. 20 Oct 2012

13. RapidMiner, free Java data mining software (2012) www.rapidminer.com. 20 Oct 2012

14. Orange, free data mining software (2012) www.orangesoftware.nu. 20 Oct 2012

15. Web G (2000) Efficient search for association rules. In: Proceedings of the 6th international conference on knowledge discovery and data mining, Boston, MA, USA, pp 99–107

16. Tan P, Kumar V, Srivastava J (2004) Get the right objective measure for association analysis. Inf Syst 29:293–313

Chapter 12
Multidimensional Scaling

1 Summary

1.1 Background

To individual patients, objective criteria of drug efficacy, like pharmaco-dynamic/-kinetic and safety measures may not mean too much, and patients' personal opinions are important too. Particularly, a drug's experienced strength, its duration of action, and side effects are important to patients.

1.2 Objective

To assess whether multidimensional scaling can visualize subgroup differences in experienced drug efficacies. To assess whether the data-based dimensions can be used to match the dimensions as expected from the pharmacological properties.

1.3 Methods

Using SPSS statistical software (PROXSCAL en PREFSCAL), we analyzed two simulated datasets: 20 patients were supposed to judge the proximities of 14 painkillers, 42 patients were supposed to scale 15 different painkillers for preference.

T.J. Cleophas and A.H. Zwinderman, *Machine Learning in Medicine:*
Part Two, DOI 10.1007/978-94-007-6886-4_12,
© Springer Science+Business Media Dordrecht 2013

1.4   Results

PROXCAL analysis was able to identify patients' priorities along both an x- and a y-axis. PREFSCAL was able to identify three clusters, consistent with patients' preferences along an x-, y-, and z-axis.

1.5   Conclusions

1. The strength of multidimensional scaling for assessing patients' opinions about drug efficacies, is, that it does not require counted estimates of patients' opinions, but rather uses estimated proximities as surrogate for that purpose.
2. The examples in the current paper suggest that proximities can, indeed, be used as estimators of patients' priorities and preferences.
3. This may sound speculative, but if the pharmacological properties of the drugs match the place of the medicines in a particular dimension, then we will be more convinced that the multi-dimensional display gives, indeed, an important insight in the real priorities of the patients.
4. Multidimensional scaling can, like regression analysis, be used both for estimating preferences/priorities in populations, and in individual patients.

2   Introduction

Drug efficacy is, usually, interpreted in terms of their pharmaco-dynamic/-kinetic properties and safety. However, to individual patients, these criteria may not mean too much, and patients' personal opinions may be important too. Multidimensional scaling is very suitable for the assessment of personal preferences in daily life, and it might, therefore, also be able to visualize subgroup differences in experienced drug efficacies. It is a computationally intensive method for information visualization of similarities and dissimilarities in datasets, first described by Samuel Torgerson [1], a bank director from Madison, Wisconsin, in 1958, but was little used, because statistical software was lacking. This changed, when it was implemented in the early versions of SPSS, developed in the 1970s at the University of California Los Angeles (UCLA) [2]. Like other machine learning methods, including factor analysis and partial least squares, it is, nowadays, widely applied in fields like econometry, sociometry, geology, market studies, etc. [3, 4]. But, in clinical research, it is virtually unused, despite the well-recognized need for better evaluation tools of treatment efficacies [5]. When searching Medline using the search term multidimensional scaling, we only found two psychiatric studies [6, 7], one meta-analysis of clinical trials [8], one public health study [9], two psychological papers

[10, 11], and, finally, one study evaluating patients' perception on drugs in general [12], and one on inhalator abuse [13]. But no systematic assessments of drug efficacies have been published to date.

The current chapter, using simulated data, studies the potential of multidimensional scaling for assessing patients' own observations about their perceived drug preferences in a systematic way. We will show step by step analyses for the convenience of investigators, and hope that this chapter will serve as a hand-hold presentation accessible to clinicians and a must-read paper to those new to the method.

3 Some Theory

As an example, we may be interested to understand patients' preferences of different painkillers. These medicines may be different in pharmacological properties like potency, plasma halftime, and other features, but this is not, what patients really care about. They just want to feel good, to get rid of their pain, few side effects, and may be other things. Personal endpoints like those are, notoriously, difficult to count. The strength of multidimensional scaling is that it does not use counted estimates of such endpoints, but rather uses estimated proximities between drugs as surrogate for that purpose. Multidimensional scaling offers two methods, proximity scaling (PROXSCAL in SPSS), and preference scaling (PREFSCAL in SPSS) [14].

3.1 Proximity Scaling

The principles of proximity scaling is shown in the example of Table 12.1. To 20 patients 14 different drugs are administered, and the patients are requested to judge the similarities between them one by one. Similarity scores scaled from 10 to 1 are used and for the analysis transformed into distance scores (1–10). For the analysis the mean scores of the 20 patients are applied. The Table 12.1 shows a matrix of these mean scores, which can be considered as one by one distances between all of the treatments. They are, subsequenty, modeled in a two-dimensional plane using the following mathematical equation (based on Pythagoras' equation for rectangular triangles).

$$\text{Distance between drug i and drug j} = \sqrt{\left[\left(x_i - x_j\right)^2 + \left(y_i - y_j\right)^2\right]}$$

All of the medicines should be connected with one another by straight lines in 14 different ways, but, unfortunately, the solutions produced by the above equation do not entirely fit the two-dimensional model, and sometimes produce wrong distances.

Table 12.1 One by one proximities between 14 pain-killers, mean estimates of 20 patients

	1	2	3	4	5	6	7	8	9	10	11	12	13	14
1	0													
2	8	0												
3	7	2	0											
4	5	4	5	0										
5	8	5	4	6	0									
6	7	5	6	6	8	0								
7	4	5	6	3	7	4	0							
8	8	5	4	6	3	8	7	0						
9	3	7	9	4	8	7	5	8	0					
10	5	6	7	6	9	4	4	9	6	0				
11	9	5	4	6	3	8	7	3	8	9	0			
12	9	4	3	7	5	7	7	5	8	9	50			
13	4	6	6	3	7	5	4	8	4	5	7	7	0	
14	6	6	7	6	8	2	4	9	7	3	9	7	5	0

This is, because the distances are mean values of experimental data, and such data come with amounts of uncertainty, normally called standard errors, or, with multidimensional scaling, called raw stress.

Raw stress = standard error = expected distance − observed distance

It is, mathematically, hard to solve the above equation, and, instead, statistical software uses iterative numerical methods to find a single solution for all of the distances with the smallest overall standard error, the best fit solution for the data given. The model is assumed to appropriately describe the data, if the stress values are < 0.20 and the dispersion values are approximately 1.0. Also goodness of fit of the modeled data is assessed by plotting the actual and fitted distances against one another. A perfect fit should produce a straight line, a poor fit produces a lot of spread around a line or even no line at all.

An important aspect of presenting proximities between the treatments in a two-dimensional way along an x- and y-axis is their orthogonal relationship. This property is also used in factor analysis, and it can be shown, that, with an orthogonal relationship between two drug effects, the magnitude of their covariance is zero, and does not have to be taken into account. It would mean, that two drug effects can be displayed by a two-dimensional scaling model, if they are (1) very important to patients and (2) largely independent of one another. In the given example both perceived pain-killing efficacy and duration of action may be assumed to be two such effects. This may sound speculative, but if the pharmacological properties of the drugs match the place of the medicines along the x- and y-axis, then we will be more convinced that the two-dimensional display gives, indeed, an important insight in the real priorities of the patients.

We should add, that, just like with the other orthogonal methodologies, the two-dimensional models can be, easily, extended to multidimensional modeling, where it is possible to distinguish between more than two experienced patients' effects. A limitation of the multiple modeling is, of course, that power is lost, that finding an appropriate best fit solution for your data becomes harder, and that the interpretation of the result is sometimes more difficult.

3.2   Preference Scaling

The principles of preference scaling are illustrated in the example of Table 12.2. To 42 patients 15 different painkillers are administered, and the patients are requested to rank them in order of preference from 1 "most prefered" to 15 "least prefered". The table shows 1 row per patient, with the 15 columns representing 1 drug per column. Like with the above similarity assessment, preference assessments can be mapped in a two dimensional plane with the rank orders of the medicines as measures of distance between the medicines. Two types of maps are possible: an aggregate map giving average distances of the entire population or individual maps of single patients.

The average maps can be interpreted very much like the above proximity scaling maps, but the individual maps can be used for another interesting purpose. Using the same mathematical equations as used above, not only a best fit map for individual patients can be constructed, but also an *ideal point* can be identified, defined as the point in an individual map with the smallest distance to all of the medicines [15]. Such a point can be interpreted as the place with an ideal medicine for a particular patient. We should add, that, mathematically, it is not always possible to find an ideal point, but, nonetheless, if the calculated ideal points as provided in the analysis are mapped together, they will produce a pattern of the patients' preferences regarding the types of medicines under investigation.

4   Data Analysis Using SPSS

4.1   Proximity Scaling

Twenty patients are requested to judge the similarities of 14 pain-killers. The Table 12.1 gives the matrix of the average one by one distance scores between the medicines. If some drugs were judged to be more similar to one another than others, then the proximities in the table should not be equally distributed. And so, we made sure that high proximities were overrepresented in the right half (horizontal dimension) and in the lower half (vertical dimension) of the table (each by 10 %).

Table 12.2 Preference orders of 15 painkillers in 42 patients (1 row = 1 patient)

Painkiller no.

1	2	3	4	5	6	7	8	9	10	11	12	13	14	15
12	13	7	4	5	2	8	10	11	14	3	1	6	9	15
14	11	6	3	10	4	15	8	9	12	7	1	5	2	13
13	10	12	14	3	2	9	8	7	11	1	6	4	5	15
7	14	11	3	6	8	12	10	9	15	4	1	2	5	13
14	9	6	15	13	2	11	8	7	10	12	1	3	4	5
9	11	15	4	7	6	14	10	8	12	5	2	3	1	13
9	14	5	6	8	4	13	11	12	15	7	2	1	3	10
15	10	12	6	8	2	13	9	7	11	3	1	5	4	14
13	12	2	4	5	8	10	11	3	15	7	9	6	1	14
15	13	10	7	6	4	9	11	12	14	5	2	8	1	3
9	2	4	13	8	5	1	10	6	7	11	15	14	12	3
11	1	2	15	12	3	4	8	7	14	10	9	13	5	6
12	1	14	4	5	6	11	13	2	15	9	3	10	8	7
11	12	13	5	4	14	10	8	7	15	3	2	6	1	9
12	11	8	1	4	7	13	10	9	14	5	2	6	3	15
15	12	3	14	5	4	10	9	7	13	6	8	1	2	11
7	10	8	3	13	6	15	12	11	9	5	1	4	2	14
7	12	6	4	10	1	15	9	8	13	5	3	14	2	11
2	9	8	5	15	12	7	10	6	11	1	3	4	13	14
10	11	15	6	9	4	14	2	13	12	8	1	3	7	5
12	1	2	10	3	15	4	6	5	13	7	11	8	9	14
13	12	10	1	11	5	15	8	7	14	2	6	4	3	9
12	6	1	14	2	5	15	8	4	13	7	10	9	3	11
10	11	9	14	5	6	12	1	3	13	8	2	11	4	7
14	8	7	5	9	11	13	3	10	6	2	1	12	4	15
12	15	8	5	9	7	14	13	11	6	4	1	3	2	10
10	3	6	14	1	7	9	4	2	5	11	15	13	12	8
6	15	3	11	9	2	13	8	10	14	5	7	12	1	4
15	7	10	2	12	9	13	8	5	6	11	1	3	4	14
14	10	6	2	9	7	15	12	8	11	5	3	1	4	13
10	4	11	9	15	8	6	5	1	13	14	2	12	3	7
9	3	10	13	14	11	2	1	4	5	15	6	7	8	12
14	8	1	11	10	2	4	13	15	9	6	5	12	3	7
14	8	3	11	10	2	4	13	15	9	6	5	12	1	7
15	6	10	14	13	8	2	4	3	5	11	1	12	7	9
12	2	13	11	9	15	3	1	4	5	6	8	10	7	14
5	1	6	11	12	10	7	4	3	2	13	9	8	14	15
15	11	7	13	4	6	9	14	8	12	1	10	3	2	5
6	1	12	5	15	9	2	7	11	3	8	10	4	14	13
15	1	5	14	4	6	3	8	9	2	12	11	13	10	7
10	3	2	14	9	1	8	12	13	4	11	5	15	6	7
13	3	1	14	4	10	5	15	6	2	11	7	12	8	9

Table 12.3 The stress (standard error) and fit measures. The best fit distances as estimated by the model are adequate: measures of stress < 0.20, values of dispersion measures (Dispersion Accounted For) and Tucker's tests with values close to 1.0

Stress and fit measures	
Normalized raw stress	.00819
Stress-I	.09051[a]
Stress-II	.21640[a]
S-stress	.02301[b]
Dispersion Accounted For (D.A.F.)	.99181
Tucker's coefficient of congruence	.99590

PROXSCAL minimizes normalized raw stress
[a]Optimal scaling factor = 1.008
[b]Optimal scaling factor = .995

SPSS statistical software [14] was be used for modeling a two dimensional plane with the best fit solution for the data as given.

Command: Analyze....Scale....Multidimensional scaling (PROXSCAL)....Data Format: click The data are proximities....Number of Sources: click One matrix source....One Source: click The proximities are in a matrix across columns....click Define....enter all variables (medicines) into "Proximities"....Model: Shape: click Lower-triangular matrix....Proximity Transformation: click Interval....Dimensions: Minimum: enter 2....Maximum: enter 2....click Continue....click Plots....mark Common space....mark Transformed proximities vs distances....click Continue....click: Output....mark Common space coordinates....mark Multiple stress measures....click Continue....click OK.

Table 12.3 gives the stress (standard error) and fit measures. The best fit distances as estimated by the model are adequate: measures of stress < 0.20, values of dispersion measures (Dispersion Accounted For) and Tucker's tests with values close to 1.0. Figure 12.1 gives a plot of the actual distances as observed versus the distances fitted by the statistical program. A perfect fit should produce a straight line, a poor fit produces a lot of spread around a line or even no line at all. The figure is not perfect but it shows a very good fit as expected from the stress and fit measures. Finally, the Fig. 12.2 shows the most important part of the outcome. The standardized x- and y-axes values give some insight in the relative position of the medicines according to perception of our study population. Four clusters were identified. The cluster at the upper right quadrant comprises high priorities of the patients along both the x- and the y-axis. The cluster at the lower left quadrant comprises low priorities of the patients along both axes. If, pharmacologically, the drugs in the right upper quadrant were highly potent with little side effects, then the patients' priorities would fairly match the pharmacological properties of the medicines.

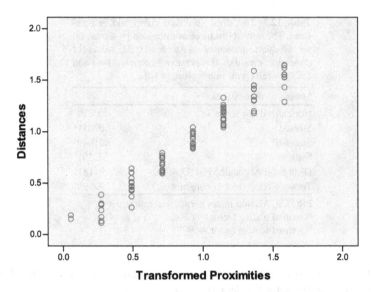

Fig. 12.1 Plot of the actual distances as observed versus the distances fitted by the statistical program. A perfect fit should produce a straight line, a poor fit produces a lot of spread around a line or even no line at all. This graph is not perfect, but it shows a very good fit as expected from the stress and fit measures (Table 12.3)

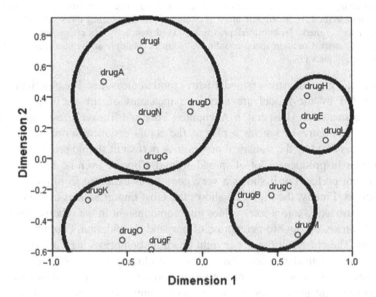

Fig. 12.2 The best fit outcome of two dimensional proximity scaling of the data from Table 12.1. The cluster at the *upper right quadrant* comprises high patients' priorities along both the x- and the y-axis. In contrast, the *left lower quadrant* would comprise low priorities along both axes

Fig. 12.3 The data from Table 12.2. To 42 patients 15 different drugs are administered, and the patients are requested to rank them in order of preference from 1 "most preferred" to 15 "least preferred". A three dimensional view of the individually assigned preferences shows a very irregular pattern consistent of multiple areas with either high or low preference

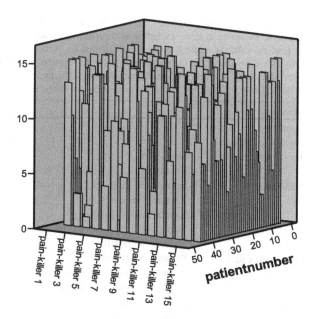

4.2 Preference Scaling

To 42 patients 15 different painkillers are administered, and the patients are requested to rank them in order of preference from 1 "most prefered" to 15 "least prefered". The Table 12.2 shows 1 row per patient, with the 15 columns representing 1 drug per column. If some drugs were judged to be more often prefered than others, then the preferences in the table should not be equally distributed. So we made sure that the pattern of distribution was irregular. Figure 12.3 gives a three dimensional view of the individually assigned preferences, and shows a very irregular pattern consisting of multiple areas with either high or low preference.

Like with the above similarity assessment, also preference assessments can be mapped in a two dimensional plane with the rank orders of the medicines as measures of distance between the medicines. Two types of maps are constructed: an aggregate map giving average distances of the entire population or individual maps of single patients, and an ideal point map where ideal points have to be interpreted as a map with ideal medicines, one for each patient. SPSS [14] is used once more.

Command: Analyze....Scale....Multidimensional Unfolding (PREFSCAL)....enter all variables (medicines) into "Proximities"....click Model....click Dissimilarities.... Dimensions: Minimum enter 2 ....Maximum enter 2....Proximity Transformations: click Ordinal ....click Within each row separately....click Continue....click Options: imputation by: enter Spearman....click Continue....click Plots: mark Final common space....click Continue....click Output: mark Fit measures ....mark Final common space....click Continue....click OK.

Table 12.4 The stress (standard error) and fit measures of the data from Table 12.2. They are adequate: measures of stress including normalized stress and Kruskal's stress-I are close to 0.20 or less, the value of dispersion measures (Dispersion Accounted For) is close to 1.0. The table also shows whether there the risk of a *degenerate* solution, which is a tendency for the average treatment places to center in the middle of the map. The DeSarbo's and Shepard criteria are close to respectively 0 % and 80 %, and we, thus, need not worry, no penalty adjustment is required

Measures		
Iterations		115
Final function value		.7104127
Function value parts	Stress part	.2563298
	Penalty part	1.9688939
Badness of fit	Normalized stress	.0651568
	Kruskal's stress-I	.2552582
	Kruskal's stress-II	.6430926
	Young's S-stress-I	.3653360
	Young's S-stress-II	.5405226
Goodness of fit	Dispersion Accounted For	.9348432
	Variance Accounted For	.7375011
	Recovered preference orders	.7804989
	Spearman's rho	.8109694
	Kendall's tau-b	.6816390
Variation coefficients	Variation proximities	.5690984
	Variation transformed proximities	.5995274
	Variation distances	.4674236
Degeneracy indices	Sum-of-squares of DeSarbo's intermixedness indices	.2677061
	Shepard's rough nondegeneracy index	.7859410

Table 12.4 gives the stress (standard error) and fit measures. The best fit distances as estimated by the model are adequate: measures of stress including normalized stress and Kruskal's stress-I are close to 0.20 or less, the value of dispersion measures (Dispersion Accounted For) is close to 1.0. The table also shows whether there is a risk of a *degenerate* solution, otherwise called loss function. The individual proximities have a tendency to form circles, and when averaged for obtaining average proximities, there is a tendency for the average treatment places to center in the middle of the map. The solution is a penalty term, but in our example we need not worry. The DeSarbo's and Shepard criteria are close to respectively 0 % and 80 %, and no penalty adjustment is required.

Figure 12.4 gives the most important part of the outcome. The standardized x- and y-axes values of the upper graph give some insight in the relative position of the medicines according to the perception of our study population. The results can be understood as the relative position of the medicines according to the perception of our study population. Both the horizontal and the vertical dimension appears to discriminate between different preferences. The lower graph gives the patients' *ideal points*. The patients were clearly split into two clusters with different preferences, although with much variation along the y-axis. The dense cluster in the right lower quadrant represented patients with preferences both along the x- and y-axis. The

Fig. 12.4 Two-dimensional preference scaling of the data from Table 12.2. *Upper graph*: the relative position of the medicines according to the perception of our study population. Both the horizontal and the vertical dimension appear to discriminate between different preferences. *Lower graph*: the patients' *ideal points*. The patients were clearly split into two clusters along the x-axis with different preferences, although with much variation along the y-axis. The dense cluster in the *right lower quadrant* represented patients with preferences both along the x- and y-axis

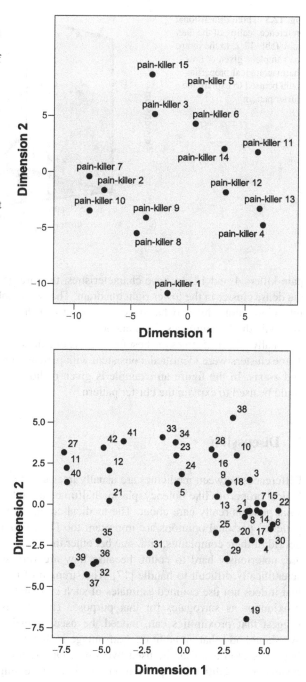

Fig. 12.5 Three-dimensional preference scaling of the data from Table 12.2. In the figure an example is given of how pharmacological properties could be used to explain the cluster pattern

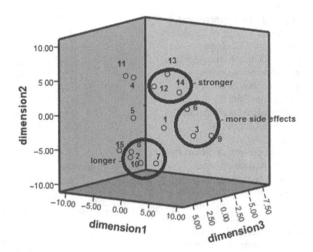

pain-killers 4 and 12–14 have characteristics, that are particularly appreciated by the dense cluster in the lower right quadrant. The two graphs are sometimes plotted into a joint plot. This may be convenient, because it shows general drug preferences and individual preferences simultaneously.

Finally, Fig. 12.5 gives the best fit outcome of a three-dimensional scaling model. Three clusters were identified, consistent with patients' preferences along an x-, y-, and z-axis. In the figure an example is given of how pharmacological properties could be used to explain the cluster pattern.

5 Discussion

Differences between medicines are usually interpreted as differences in pharmacological properties like potency, plasma halftime, and other features, but this is not, what patients really care about. The medical community has come to realize that patients' personal opinions are important too [16]. Patients want to feel good, to get rid of their complaints, and, maybe, other things. Personal endpoints like these are, notoriously hard to count, because they are very subjective, and, therefore, scientifically difficult to handle [17]. The strength of multidimensional scaling is, that it does not use counted estimates of such endpoints, but rather uses estimated proximities as surrogates for that purpose. The examples in the current paper suggest that proximities can, indeed, be used as estimators of drug efficacies. We should add that an additional advantage is that multidimensional scaling can, like regression analysis, be used two ways. One is for estimating preferences of treatment modalities in a population, one is for assessing the preferred treatment modalities in individual patients.

Limitations have to mentioned. First, proximities may have different meanings to different patients: in one patient it may indicate lack of side effects, in the other high strength etc. However, multidimensional scaling enables to assess multiple dimensions each of which can be assigned to one particular cause for proximity. This may sound speculative, but if the pharmacological properties of the drugs match the place of the medicines in a particular dimension, then we will be more convinced that the multi-dimensional display gives, indeed, an important insight in the real priorities of the patients. Second, generally, the measured proximities do not entirely fit the model and an approximate rather than exact solution is offered. This may enhance the chance of wrong interpretations of the data. Third, wrong interpretations may also follow from a situation where all proximities are approximately equal. It would mean no preferences observed. However, the model will produce alls positions on concentric circles. Fourth, a patient's preference may have a single reason, e.g., high strength. Multidimensional scaling is not appropriate and will produce a bad fit for patients with single reasons. Other limitations are that personal opinions may be biased, that personal opinions are perceived data, not real (objectively measured) data, that are somewhat comparable with the latent variables of factor analysis.

6 Conclusions

Despite the above limitations, the method does help both to underscore the pharmacological properties of the medicines under studies, and to identify what effects are really important to patients. We conclude.

1. The strength of multidimensional scaling for assessing patients' opinions about drug efficacies, is, that it does not require counted estimates of patients' opinions, but rather uses estimated proximities as surrogate for that purpose.
2. The examples in the current paper suggest that proximities can, indeed, be used as estimators of drug efficacies.
3. An additional advantage is that multidimensional scaling can, like regression analysis, be used two ways, (1) for estimating preferences of treatment modalities in a population, (2) for assessing the preferred treatment modalities in individual patients.
4. When the data-based dimensions match the dimensions as expected from the pharmacological properties, the method helps to underscore the pharmacological properties, and to identify what effects are really important to patients.

References

1. Torgerson WS (1958) Theory and methods of scaling. Wiley, New York
2. SPSS (2012) www.en.wikipedia.org/wiki/SPSS. 10 Nov 2012
3. Groenen P, Van de Velden M (2004) Multidimensional scaling. Economic Institute Report EI 2004-15. Erasmus University Rotterdam, the Netherlands

4. Takane Y, Jung S, Oshima-Takane Y (2012) Multidimensional scaling. www.takane.brinkster. net/yoshio/c043.pdf. 10 Nov 2012
5. US Department of Health and Human Services and Food & Drug Administration (2006) Better evaluation tools. In: Critical path opportunities list. March, pp 1–8, www.fda.gov. 10 Nov 2012
6. Carlier I, Lamberts R, Gersons B (2000) The dimensionality of trauma: a multidimensional scaling comparison of police officers with or without posttraumatic stress disorder. Psychiatry Res 97:29–39
7. Nakao K, Miulazzo L, Rosenstein M, Manderscheid R (1986) Referral patterns to and from in patients psychiatric services: a social network approach. Am J Public Health 76:755–760
8. Chung H, Lumley T (2008) Graphical exploration of network meta-analysis data: the use of multidimensional scaling. Clin Trials 5:301–307
9. Bartolucci AA (1986) Multidimensional scaling and the information it conveys. Am J Public Health 76:747–748
10. Smith JK, Falvo D, McKilip J, Pitz G (1984) Measuring patient perceptions of patient-doctor interaction. Eval Health Prof 7:77–94
11. Dew PB (1991) Extrapolation by proximate inference. Neurosci Biobehav Res 15:149–151
12. Kessler RC, Paton SM, Kandel RB (1976) Reconciliation unidimensional and multidimensional models of patterns of drug use. J Stud Alcohol 37:632–647
13. Frank B (1995) Gathering epidemiological information on inhalant abuse: some methodological issues. NIDA Res Monogr 148:260–273
14. SPSS statistical software (2012) www.spss.com. 10 Nov 2012
15. Huber J (2012) Ideal point models of preference. www.acrwebsite.org. 31 Oct 2012
16. McPherson K (2009) Do patients'preferences matter? BMJ 338:59–60
17. Testa MA, Simonson DC (1996) Assessment of quality of life outcomes. N Engl J Med 334:835–840

Chapter 13
Correspondence Analysis

1 Summary

1.1 Background

Multiple treatments for one condition are increasingly available, and a systematic assessment would serve optimal care. Research in this field to date is problematic.

1.2 Objective

To propose a novel method based on cross-tables, correspondence analysis.

1.3 Methods

We used a simulated example of three treatment groups and three response groups. SPSS statistical software was applied for correspondence analysis.

1.4 Results

A significant difference in response patterns between the treatment groups was observed, chi-square 21.462, $p < 0.0001$. The correspondence analysis revealed that two of the treatment groups tended to join two of the response groups. A 2×2 table showed that the observed tendencies were statistically significant, chi-square 11.9, $p < 0.0001$.

T.J. Cleophas and A.H. Zwinderman, *Machine Learning in Medicine:*
Part Two, DOI 10.1007/978-94-007-6886-4_13,
© Springer Science+Business Media Dordrecht 2013

Table 13.1 Cross-table of treatments by response group. Both the observed counts (patients) and the expected counts, if no significant difference between the cells existed

Treatment by remission cross-tabulation

			Remission, partial, no			
			1.00	2.00	3.00	Total
Treatment	1.00	Count	19	21	18	58
		Expected count	21.6	10.7	25.7	58.0
	2.00	Count	41	9	39	89
		Expected count	33.2	16.4	39.4	89.0
	3.00	Count	21	10	39	70
		Expected count	26.1	12.9	31.0	70.0
Total		Count	81	40	96	217
		Expected count	81.0	40.0	96.0	217.0

1.5 Conclusions

1. Correspondence analysis is adequate for comparing categories, but has not yet been used for comparing treatment groups with different treatments, although such groups can, conveniently, be classified as sets of categories.
2. In the example given it was able to demonstrate in an unbiased way which of the treatments was best.
3. Traditional tests suffered from the bias of chance findings due to multiple testing.

2 Introduction

Multiple treatments for one condition are increasingly available, and a systematic assessment would serve optimal care [1–3]. However, research in this field is problematic. Multiple treatments studies tended to systematically overestimate the treatment effects [1]. Direct comparisons using Kaplan-Meier methodology suffered from confounding [2], and propensity scores as a solution for this flaw has not been entirely accepted by the scientific community [4]. Also, multiple treatments require multiple comparisons, that tend to be complex and not very transparent [3]. In the current chapter we propose a novel method based on cross-tabulations.

Cross-tables of two sets of categorical data is a very simple and common form of analysis in research. The Table 13.1 shows a cross-table of three treatment groups versus three response groups ((1) remission, (2) partial remission, and (3) no remission). The number of patients in each cell are given as counts. Treatment 1 is dominant in response group 2, treatment 2 in response group 1, and both the treatments 2 and 3 are dominant in response group 3.

In order to test whether there is a significantly different pattern in numbers of responders between the different treatment groups, a multiple groups chi-square test is performed. It compares the observed counts per cell with the expected number, if no significant difference between the cells existed. The chi-square value equals 21.462 with 4 degrees of freedom, p-value < 0.0001. There is a significant difference in response patterns between the treatment groups. In order to find out which treatment performs best, nine different 2×2 tables, as constructed from the nine cells, can be analyzed. But this procedure raises the problem of multiple testing, and false positive findings due to type I errors of finding a difference where there is none. In order to avoid this potential bias, Jean Paul Benzecri, statistician at Sorbonne University Paris in 1973 [5] developed a multivariate method, conceptually similar to multivariate analysis of variance, but using categories, rather than continuous data. The convenience of the method is, that it provides a map of proximities between the members of two sets of categories, and the members with the closest proximities can, subsequently, be tested for statistical significance. The method is, computationally, somewhat laborious, but, fortunately, available in many software programs including SPSS statistical software [6]. It is widely applied in marketing research as an important machine learning technology for predicting product sales [7, 8]. In clinical research it is little used. When searching Medline we found one health care marketing study [9], one epidemiological study [10], one genetic study [11], and four diagnostic studies [12–15], but no therapeutic studies.

The current chapter, using a simulated example, assesses whether correspondence analysis is suitable for therapeutic studies comparing multiple treatments, and whether it is able to avoid the problems associated with the cross-tab analyses as traditionally used for the purpose.

We will use step by step analyses for the convenience of the readers. This is a hand-hold presentation for clinicians and must-read chapter for those new to the method.

3   Some Theory

As an alternative to multiple 2×2 tables for finding the best treatment in a multiple treatments assessment, a correspondence analysis can be performed. Correspondence analysis is based on the principle that the coordinates of 3 % that add up to 100 % plotted in a three-dimensional space, are, actually, in a two-dimensional plane. Figure 13.1 gives a simple example of this principle. A three-axes model with rectangular sides of length = 100 % is drawn. If the rectangle in the basal plane with sides of 25 % and 35 % is moved up by 40 % (arrow), then it should touch the frontal plane, because 25 % + 35 % + 40 % = 100 %. In general, any coordinate of three axes should be in the frontal plane as long as the coordinates add up to 100 %.

This means that such coordinates can, indeed, be displayed in a two-dimensional plane.

Therefore, if, in a given example, you plot the coordinates of three treatment groups in this way, using the percentages given by the response groups (Table 13.2 upper part), then you will observe that all of the coordinates are in one and the

Fig. 13.1 Three-axes model with an add-up sum of the rectangular sides of the length of 100 %. If the rectangle in the basal plane with sides of 25 % and 35 % is moved up by 40 % (*arrow*), then it should touch the frontal plane. This is, because 25 % + 35 % + 40 % = 100 %. In general, any coordinate of three axes should be in the frontal plane as long as the coordinates add up to 100 %

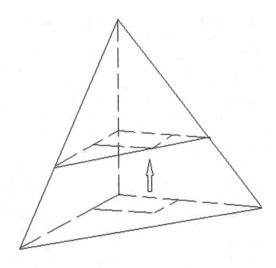

Table 13.2 Row and column profiles of the data from Table 13.1

Response group				
Treatment	(1) Remission	(2) Partial	(3) No	Total
1	32.8 %	36.2 %	31.0 %	100 %
2	46.1 %	10.1 %	43.8 %	100 %
3	30.0 %	14.3 %	55.7 %	100 %
Total	37.3 %	18.4 %	44.2 %	100 %

Treatment group				
Response	1	2	3	Total
1 (Remission)	23.5 %	50.6 %	25.9 %	100 %
2 (Partial)	52.5 %	22.5 %	25.0 %	100 %
3 (No)	18.8 %	40.6 %	40.6 %	100 %
Total	26.7 %	41.0 %	21.7 %	100 %

same plane. Similarly, if you plot the coordinates of three response groups in this way, using the percentages given by the treatment groups (Table 13.2 lower part), then the three coordinates will be in the same plane. This plane can, subsequently, be used to visualize whether treatment groups and response groups tend to join or not. Mapping the coordinates of the treatment groups and response groups in this way is meaningful. If, for example, treatment group 2 and response group 1 tend to join, and if response group 1 means complete remission, then we have reason to believe that treatment group 2 has the best treatment. If, for example, treatment group 1 tends to join with response group 2 (partial remission), then treatment group 1 may have the second best treatment. A 2 × 2 table can be used to test whether the observed tendencies are statistically significant.

Table 13.3 Table of similarity values as calculated from the nine cells of Table 13.1

Treatment		Remission		
		Yes	Partial	No
1	Residual	−2.6	10.3	−7.7
	$(o-e)^2/e$	0.31	9.91	2.31
	Similarity	−0.31	9.91	−2.31
2	Residual	7.9	−7.4	−0.4
	$(o-e)^2/e$	1.88	3.34	0.004
	Similarity	1.88	−3.34	−0.004
3	Residual	−4.1	−2.9	8.0
	$(o-e)^2/e$	0.64	0.65	2.65
	Similarity	−0.64	−0.65	2.65

4 Example

In the Appendix is the data file of the example used in the chapter: 217 patients were randomly treated with one of three treatments (treat = treatment) and produced one of three responses (1 = complete remission, 2 = partial remission, 3 = no response). We will use SPSS statistical software [6]. First, a multiple groups chi-square test is performed.

Command: Analyze ....Descriptive Statistics ....Crosstabs ....Row(s): enter treatment .... Column(s): enter remission, partial, no [Var 2]....click Statistics....mark Chi-square ....click Continue ....click Cell Display....mark Observed ....mark Expected .... click Continue ....OK.

Table 13.1 is in the outcome file. It compares the observed counts (patients) per cell with the expected count, if no significant difference existed. Also, a chi-square value is given, 21.462 with 4 degrees of freedom, p-value < 0.0001. There is a significantly different pattern in numbers of responders between the different treatment groups. To find out what treatment is best a correspondence analysis is performed. For that purpose the individual chi-square values are calculated from the values of Table 13.1 according to the underneath equation.

$$\left[(\text{observed count} - \text{expected count})^2 / \text{expected count} \right]$$

Then, the individual chi-square values are converted to similarity measures as demonstrated in Table 13.3. With these values the software program creates a two-dimensional quantitative distance measure that is used to interpret the level of nearness between the treatment groups and response groups. We will use again SPSS statistical software for the analysis.

Command: Analyze ....Dimension Reduction ....Correspondence Analysis ....Row: enter treatment ....click Define Range ....Minimum value: enter1 ....Maximum value: enter 3 ....click Update ....Column: enter remission, partial, no [Var 2] ....click

Fig. 13.2 A plot of the coordinates of both the treatment groups and the response groups in one two-dimensional plane is meaningful. As treatment group 2 and response group 1 (complete remission) tend to join, and treatment group 1 and response group 2 (partial remission) do so, we have reason to believe that treatment group 2 may have the best treatment, and treatment group 2 the second best

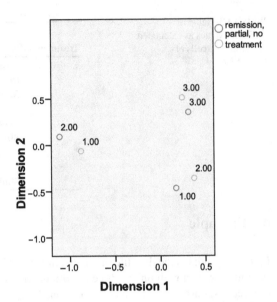

Define Range....Minimum value: enter1....Maximum value: enter 3....click Update....click Continue....click Model....Distance Measure: click Chi square....click Continue....click Plots....mark Biplot....OK.

A plot of the coordinates of both the treatment groups and the response groups in a two-dimensional plane is shown in Fig. 13.2. This plot is meaningful. As treatment group 2 and response group 1 tend to join, and treatment group 1 and response group 2 do, equally, so, we have reason to believe that treatment group 2 has the best treatment and treatment group 1 the second best. This is, because response group 1 has a complete remission, and response group 2 has a partial remission. If a 2×2 table of the treatment groups 1 and 2 versus the response groups 1 and 2 shows a significant difference between the treatments, then we can argue, that the best treatment is, indeed, significantly better than the second best treatment.

	Response		
Treatment	1	2	Total
1	19	21	40
2	41	9	50
	60	30	90

For the analysis recoding of the variables is required, but a simpler solution is to use a pocket calculator method for computing the chi-square value.

$$\text{Chi-square} = \frac{[(9 \times 19) - (21 \times 41)]^2 \times 90}{60 \times 30 \times 50 \times 40} = 11.9 \text{ with 1 degree of freedom,}$$

$$p < 0.0001$$

Treatment 2, indeed, produced significantly more complete remissions than did treatment 1, as compared to the partial remissions.

5 Discussion

Multiple treatments for one condition are increasingly available [1–3]. An international working committee of the International Society of Pharmacoeconomics and Outcomes Research (ISPOR) recently stated that decisions about optimal care must rely on evidence-based evaluation and identification of the best treatment [3]. However, research in this field is problematic for at least three reasons. First, multiple treatment studies tend to overestimate the treatment effect [1]. Second, they suffer from confounders [2]. Third, they require complex multivariate statistics, causing lack of transparency [3]. Correspondence analysis, which is very adequate for comparing categories in general [7, 8], has not been used for comparing patient groups with different treatments, although such groups can, conveniently, be classified as sets of categories.

In our example correspondence analysis was able to demonstrate which one of three treatments was best, and it needed, instead of *multiple* 2×2 tables, only a single 2×2 table for that purpose. The advantage of this procedure will be even more obvious, if larger sets of categorical data have to be assessed. A nine cells data file (Table 13.1) would require only nine 2×2 tables to be tested, a 16 cells data file would require 36 of them. This procedure will almost certainly produce significant effects by chance rather than true effects, and is, therefore, rather meaningless. In contrast, very few tests are needed, when a correspondence analysis is used to identify the proximities in the data, and the risk of type I errors is virtually negligible.

We should add that, instead of a two-dimensional analysis as used in the current chapter (Figs. 13.1 and 13.2), correspondence analysis can also be applied for multidimensional analyses. For that purpose you have to enter the number of dimensions requested after the commands: Model....Dimension in solution [6]. However, with $n \times n$ tables, the maximum number of dimensions is limited to $n-1$. It is, therefore, adjusted by the software to 2, if you entered 3 or more. With larger tables, multiple correspondence analysis is readily possible, although the results are displayed in the form of tables rather than plots. A problem is that data are increasingly hard to interpret with larger tables.

6 Conclusions

1. Correspondence analysis is adequate for comparing categories but has not yet been used for comparing treatment groups with different treatments, although such groups can conveniently be classified as a set of categories.

2. In the example given it was able to demonstrate in an unbiased way which of the treatments was best.
3. Traditional tests suffered from the bias of chance findings due to multiple testing.

Appendix

Data file of the example used in the chapter: 217 patients were randomly treated with one of three treatments (treat = treatment) and produced one of three responses (1 = complete remission, 2 = partial remission, 3 = no response)

Treat	Response	Treat	Response	Treat	Response	Treat	Response	Treat	Response	Treat	Response
1	1	1	2	2	1	2	3	3	1	3	1
1	1	1	2	2	1	2	3	3	1	3	1
1	1	1	3	2	1	2	3	3	1	3	1
1	1	1	3	2	1	2	3	3	1	3	1
1	1	1	3	2	1	2	3	3	1	3	1
1	1	1	3	2	1	2	3	3	1	3	1
1	1	1	3	2	1	2	3	3	1	3	1
1	1	1	3	2	1	2	3	3	1	3	1
1	1	1	3	2	1	2	3	3	1	3	1
1	1	1	3	2	1	2	3	3	1	3	1
1	1	1	3	2	1	2	3	3	1	3	1
1	1	1	3	2	1	2	3	3	1	3	1
1	1	1	3	2	1	2	3	3	1	3	1
1	1	1	3	2	1	2	3	3	1	3	1
1	1	1	3	2	1	2	3	3	1	3	1
1	1	1	3	2	1	2	3	3	2	3	2
1	1	1	3	2	1	2	3	3	2	3	2
1	1	1	3	2	1	2	3	3	2	3	2
1	2	1	3	2	1	2	3	3	2	3	2
1	2	2	1	2	1	2	3	3	2	3	2
1	2	2	1	2	1	2	3	3	2	3	2
1	2	2	1	2	2	2	3	3	2	3	2
1	2	2	1	2	2	2	3	3	2	3	2
1	2	2	1	2	2	2	3	3	2	3	2
1	2	2	1	2	2	2	3	3	2	3	2
1	2	2	1	2	2	2	3	3	3	3	3
1	2	2	1	2	2	2	3	3	3	3	3

(continued)

(continued)

Treat	Response	Treat	Response	Treat	Response	Treat	Response	Treat	Response	Treat	Response
1	2	2	1	2	2	2	3	3	3	3	3
1	2	2	1	2	2	2	3	3	3	3	3
1	2	2	1	2	2	2	3	3	3	3	3
1	2	2	1	2	3	2	3	3	3	3	3
1	2	2	1	2	3	2	3	3	3		
1	2	2	1	2	3	3	1	3	3		
1	2	2	1	2	3	3	1	3	3		
1	2	2	1	2	3	3	1	3	3		
1	2	2	1	2	3	3	1	3	3		

References

1. Caldwell D, Ades A, Higgins J (2005) Simultaneous comparison of multiple treatments: combining direct and indirect evidence. BMJ 331:897–904
2. Sparapani R (2005) Multiple treatments, confounding and the propensity score. www.mcw.edu. 20 Nov 2012
3. Cooper N, Peters J, Lai M, Juni P, Wandel S, Palmer S, Paulden M, Conti S, Welton N, Abrams K, Bujkiewicz S, Spiegelhalter D, Sutton A (2011) How valuable are multiple treatment comparison methods in evidence-based healthcare evaluation? Value Health 14:371–380
4. Joffe M, Rosenbaum P (1999) Propensity scores. Am J Epidemiol 4:327–331
5. Benzecri JP (1973) L'analyse de donnees, Vol 2. L'analyse de correspondances. Edit by Dunod, Paris
6. SPSS statistical software. www.spss.com. 20 Nov 2012
7. Hoffman DL, Franke GR (1986) Correspondence analysis: graphical representation of categorical data in marketing research. J Market Res 23:213–227
8. Bendixen M (2003) A practical guide to the use of correspondence analysis in marketing research. Market Bull 14:1–15
9. Javalgi R, Whipple T, McManamon M, Edick V (1992) Hospital image: a correspondence analysis. J Health Care Market 12:34–41
10. Sourial N, Wolfson C, Zhu B, Quail J, Fletcher J, Karunananthan S, Bandeen-Roche K, Beland F, Bergman H (2010) Correspondence analysis is a useful tool to uncover the relationships among categorical variables. J Clin Epidemiol 63:638–646
11. Tan Q, Brusgaard K, Kruse TA, Oakeley E, Hemmings B, Beck-Nielsen H, Hansen L, Gaster M (2004) Correspondence analysis of microarray time-course data in case–control design. J Biomed Inform 37:358–365
12. Williams L (2010) A tutorial on multiblock discriminant correspondence analysis: a new method for analyzing discourse data from clinical populations. J Speech Lang Hear Res 53:1372–1393
13. Beaton D, Abdi H (2011) Partial least squares-correspondence analysis: a new method to analyze common patterns in measures of cognition and genetics. Neuroinformatics www.neuroinformatics2011.org. 20 Nov 2012
14. Almeida R, Infantosi A, Suassuna J, Costa J (2009) Multiple correspondence analysis in predictive logistic modeling: application to living-donor kidney transplant data. J Comput Method Programs Biomed 95:116–128
15. Crichton N, Hinde J (1989) Correspondence analysis as a screening method for indicants for clinical diagnosis. Stat Med 8:1351–1362

Chapter 14
Multivariate Analysis of Time Series

1 Summary

1.1 Background

Time series are encountered in every field of medicine. Traditional tests are unable to assess trends, seasonality, change points and the effects of multiple predictors like treatment modalities simultaneously. Autoregressive integrated moving average (ARIMA) is able to do all of that, but is, virtually, unused in medicine.

1.2 Objective

To assess the performance of ARIMA modeling in medicine.

1.3 Methods

A simulated time series of a population with diabetes type II was used with variables like numbers of nurse, doctor, and phone visits as predictors.

1.4 Results

A good fit of the observed data was given by the ARIMA model, as demonstrated by insignificant stationary R square and Ljung-Box tests. Three independent predictors were identified with p-values of < 0.0001. Seven significant outliers (p-values between < 0.001 and < 0.0001) explained the effects of the independent predictors.

T.J. Cleophas and A.H. Zwinderman, *Machine Learning in Medicine:*
Part Two, DOI 10.1007/978-94-007-6886-4_14,
© Springer Science+Business Media Dordrecht 2013

1.5   Conclusions

1. ARIMA modeling is very sensitive for a simultaneous assessment of trends, seasonality, change points, and the effects of multiple predictors.
2. It is a multivariate methodology, preventing the type I error risk of multiple testing as commonly observed with traditional tests.
3. It can be applied both to analyze time series in individual patients and in populations.

2   Introduction

Time series often appear as a sequence of unpaired or paired data observed through time. Examples include the incidence of nosocomial infection in a hospital, course of analgesia during surgery, seasonal variations in hospital admissions, analysis of ambulatory blood pressures, the course of any disease through time. In fact, time series are encountered in virtually every field of medicine. The analysis of time series, traditionally, focuses on the identification of patterns, and the prediction of future observations. Unlike Kaplan-Meier methodology which assesses the time to a single event in a group of patients, time series deals with multiple repeated observations and/or events either paired or unpaired.

Four specific questions are most often assessed.

1. Is there a trend in the magnitude or frequency of events through time?
2. Are there cyclic patterns in the long term data?
3. Is there a point in time at which a pattern changes?
4. Are there any predictors like treatments?

All of these questions can be simultaneously addressed by a method called the autoregressive integrated moving average (ARIMA) model. It is based on autoregression (AR), a technique invented in the early 1960s by Udney Yule [1], a Cambridge UK professor of statistics. Time curves are cut into pieces and, then, compared with one another using linear regression. In this way it can demonstrate significantly repetitive patterns, meaning patterns that are stronger repetitive than could happen by chance. However, in order to demonstrate change points from a repetitive pattern and to demonstrate upward or downward trends, the comparison of the averages of a number of observations is more sensitive. An extended method for the purpose was proposed by Box and Jenkins [2]: it is called ARMA (autoregressive moving average), and assesses autocorrelations, change points and up/downward trends simultaneously.

It would be nice, if, just like with multiple regression, independent predictors of the outcome like treatments could be included. This is accomplished by the ARIMA (autoregressive integrated moving average) method, a multivariate and computationally intensive method, available in many software programs, e.g., in

SPSS's module Forecasting [3]. ARIMA is currently widely applied in social sciences for making predictions [4], but is little used in clinical research, despite the recognized need for better predictive models in this field [5, 6]. When searching Medline ARIMA studies were found sporadically: two organisational health care studies [7, 8], two single subject studies [9, 10], and only one therapeutic study [11].

The current chapter using a simulated example assesses whether ARIMA is able to identify seasonal, trend effects and outlier effects in a longitudinal patient studies, and, at the same, is able to adjust for independent predictors of the modeled effects. Step by step analyses using SPSS statistical software are given for the benefit of investigators.

3 Some Theory

Autocorrelations is a technique that cuts time curves into pieces. These pieces are, subsequently, compared with the original datacurve using linear regression analysis. In an outpatient clinic C-reactive protein values may be higher in winter than in summer, and Fig. 14.1 gives a simulated example of the pattern of C-reactive protein values in a healthy subject. The curve is cut into pieces four times, and the cut pieces are called lagcurves, and they are moved to the left end of the original curve. The first-row lagcurve in Fig. 14.1 is very close to the original datacurve. When performing a linear regression analysis with the original data on the y-axis and the lagdata on the x-axis, a strong positive correlation will be found. The second lagcurve is not close anymore, and linear regression of the two curves produces a correlation coefficient of approximately zero. Then, the third lagcurve gives a mirror image of the original datacurve, and, thus, has a strong negative correlation. Finally, the fourth lagcurve is in phase with the original datacurve, and, thus, has a strong positive correlation.

If, instead of a few lagcurves, monthly lagcurves are produced, then we will observe that the magnitude of the autocorrelation coefficients changes sinusoidally in the event of seasonality.

Autocorrelation coefficients significantly larger or smaller than 0 must be observed in order to conclude the presence of a statistically significant autocorrelation. The underneath equation shows the mathematical equation of autocorrelation of two subsequent observations in a time series. The equation is derived from the correlation coefficient of linear regression.

$$\text{Autocorrelation coefficient (ACC)} = r = (x_t - \bar{x})\,(x_{t+1} - \bar{x})\,/\,(x_t - \bar{x})^2$$

x_t is the t-th observation in a time series and x is the mean of all observations. With one observation per time unit (day or week or month) an autocorrelation coefficient versus the first observation is calculated, and all of them are drawn along the time axis. In this way an overview of the autocorrelation function can be given.

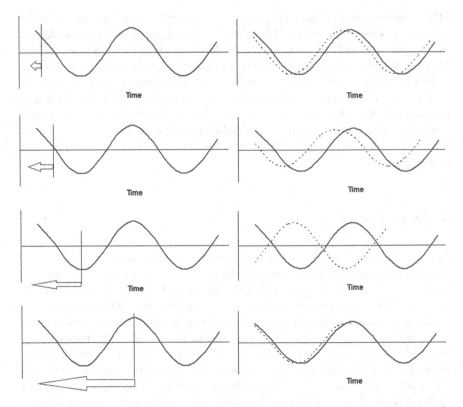

Fig. 14.1 Seasonal pattern in C-reactive protein levels in a healthy subject. Lagcurves (*dotted*) are partial copies of the datacurve moved to the left as indicated by the *arrows*. *First-row graphs*: the datacurve and the lagcurve has largely simultaneous positive and negative departures from the mean, and, thus, has a strong positive correlation with one another (correlation coefficient $\approx +0.6$). *Second-row graphs*: this lagcurve has little correlation with the datacurve anymore (correlation coefficient ≈ 0.0). *Third-row graphs*: this lagcurve has a strong negative correlation with the datacurve (correlation coefficient ≈ -1.0). *Fourth-row graphs*: this lagcurve has a strong positive correlation with the datacurve (correlation coefficient $\approx +1.0$)

Table 14.1 gives an example. We wish to find an appropriate equation for the curve of the autocorrelation coefficients in Fig. 14.2. For that purpose we, simply, use changing linear functions (a = intercept, b = direction coefficient).

$$y = a + bx$$
$$ACC_1 = ACC_0 + b_1 \, t_1$$
$$ACC_2 = ACC_0 + b_1 \, t_1 + b_2(t_2 - t_1)$$

The above equation can also be described as

$$ACC_2 = ACC_0 + (b_1 - b_2) \, t_1 + b_2(t_2)$$

Table 14.1 SPSS table presenting the autocorrelation coefficients and their standard errors. Instead of the t-test, the Ljung-Box test is used for statistical testing. This method assesses overall randomness on numbers of lags instead of randomness at each lag and is calculated according to Chi-square $= n\,(n+2)\,\Sigma\,[r_k^2/(n-k)]$ (with h degrees of freedom) with n = sample size, k = lag number, Σ = add-up sum of terms from k = 1 to k = h, with h = total number of lags tested). It produces much better p-values than does the t-statistic. However, the p-values given do not mean too much without adjustment for multiple testing. SPSS uses Hochberg's false discovery rate method. The adjusted 95 % confidence intervals are given in Fig. 14.2 lower graph

Autocorrelations

Series: VAR00008

Lag	Autocorrelation	Std. error[a]	Box-Ljung statistic		
			Value	df	Sig.[b]
1	.189	.192	.968	1	.325
2	.230	.188	2.466	2	.291
3	−.084	.183	2.678	3	.444
4	−.121	.179	3.137	4	.535
5	−.541	.174	12.736	5	.026
6	−.343	.170	16.825	6	.010
7	−.294	.165	19.999	7	.006
8	−.184	.160	21.320	8	.006
9	.080	.155	21.585	9	.010
10	.239	.150	24.141	10	.007
11	.237	.144	26.838	11	.005
12	.219	.139	29.322	12	.004
13	.343	.133	35.982	13	.001
14	.026	.127	36.024	14	.001
15	−.087	.120	36.552	15	.001
16	−.166	.113	38.707	16	.001

[a]The underlying process assumed is independence (white noise)
[b]Based on the asymptotic chi-square approximation

With multiple ACCs and the replacement of the "b−b" terms by "φ" terms, the equation will look like:

$$ACC_t = ACC_0 + \varphi_1\,t_1 + \varphi_2\,t_2 + \varphi_3\,t_3 + \varphi_4\,t_4 + \varphi_5\,t_5 + \ldots\ldots\ldots\varphi_t\,t_t$$

The above equation is called an autoregressive model of the t-th order. We wish to test whether the equation is significantly different from zero.

For that purpose standard errors are required. Instead of t-tests, the Ljung-Box tests are used. This method assesses overall randomness on numbers of lags instead of randomness at each lag and is calculated according to

$$\text{Chi-square} = n\,(n+2)\,\Sigma\,[r_k^2/(n-k)] \quad \text{(with h degrees of freedom)}$$

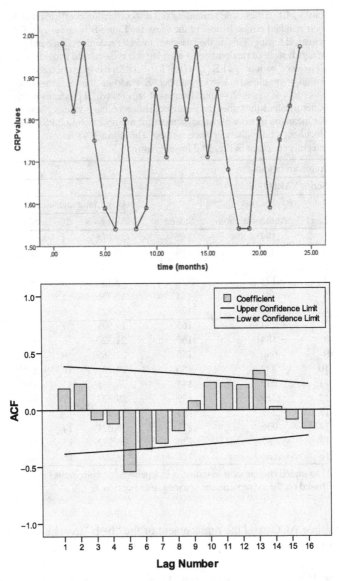

Fig. 14.2 *Upper graph*: the mean monthly CRP values (mg/l) in a healthy population look inconsistent. *Lower graph*: a sinusoidal time-pattern of the autocorrelation coefficients (*ACFs*) is observed, with a significant positive autocorrelation at the month 13 with a correlation coefficient of 0.34 (SE 0.13)

with n = sample size, k = lag number, Σ = add-up sum of terms from k = 1 to k = h, with h = total number of lags tested). It produces better p-values than does the t-statistic. However, the p-values given do not mean too much without adjustment for multiple testing. SPSS uses Hochberg's false discovery rate method. The adjusted 95 % confidence intervals are given in Fig. 14.2 lower graph. Using this approach, we can assess, whether the data are stationary or change periodically. As long as the 95 % confidence intervals are not crossed, we conclude that the AR model is stationary. At 5 and 13 months the 95 % confidence intervals are transiently crossed. However, 1 month later the stationary pattern has returned. These data are suitable for AR modeling. The best fit φ values (otherwise called parameters) are calculated just like with multiple linear regression using ordinary least squares.

In case of a more lasting crossing of the 95 % confidence, another model is more sensitive, that is the MA (moving average) model. It does not apply ACCs, but rather the actual observations and their means, but, otherwise, it also makes use of the same method of changing linear functions. It uses the following mathematical equation.

$$y = a + bx$$
$$observation_1 = observation_0 + b_1\, t_1$$
$$observation_2 = observation_0 + b_1\, t_1 + b_2(t_2 - t_1).$$

The above equation can also be described as

$$observation_2 = observation_0 + (b_1 - b_2)\, t_1 + b_2(t_2).$$

With multiple observations and the replacement of the "b-b terms" by "λ terms" the equation will look like:

$$observation_t = observation_0 + \lambda_1\, t_1 + \lambda_2\, t_2 + \lambda_3\, t_3 + \lambda_4\, t_4 + \lambda_5\, t_5 + \ldots\ldots\lambda_t\, t_t$$

The best fit λ values are calculated just like with multiple linear regression using ordinary least squares.

A more sophisticated approach is ARIMA (autoregressive integrated moving average) methodology. It is a multivariate method that finds the best fit parameters for the AR (autoregressive) part of the data and the best fit parameters for the MA (moving average) part of the data, and, at the same time, adjusts the two parts for one another. The mathematics is more computationally intensive, requiring SSCP (sum of squares and cross products) matrices and iteration methods, but excellent software is available, like the module Forecasting in SPSS.

4   Example

Monthly HbA1c levels in patients with diabetes type II are a good estimator for adequate diabetes control, and have been demonstrated to be seasonal with higher levels in the winter [12]. A large patient population was followed for 10 year. The mean values are in the data (Appendix). We will first assess the observed values along the time line. The analysis is performed using SPSS statistical software.

> Command: analyze....Forecast....Sequence Charts....Variables: enter HbA1c....Time
> Axis Labels: enter Date....OK.

Figure 14.3 upper part shows the observed data. There are (1) numerous peaks, which are (2) approximately equally sized, and (3) there is an upward trend: (2) suggests periodicity which was expected from the seasonal pattern of HbA1c values[2], (3) is also expected, it suggests increasing HbA1c after several years due to beta-cell failure. Finally (4), there are several peaks that are not part of the seasonal pattern, and could be due to outliers. Outliers can be addressed using the Expert Modeler.

> Command: Analyze....Forecast....Time Series Modeler....Dependent Variables:
> enter HbA1c....Independent Variables: enter nurse, doctor, phone, self control,
> and patient meeting....click Methods: Expert Modeler....click Criteria....Click
> Outlier Table....Select automatically....Click Statistics Table....Select Parameter
> Estimated....Click Plots table....Click Series, Observed values, Fit values....click OK.

Figure 14.3 lower graph shows that a good fit of the observed data is given by the ARIMA model, and that an adequate predictive model is provided. The upward trend is in agreement with beta-cell failure after several years. Table 14.2 shows that three significant predictors have been identified. Also the goodness of fit of the ARIMA (p, d, q) model is given, where p = number of lags, d = the trend (one upward trend means d = 1), and q = number of moving averages (= 0 here). Both Stationary R square, and Ljung-Box tests are insignificant. A significant test would have meant poor fit. In our example, there is an adequate fit, but the model has identified no less than seven outliers. Table 14.3 indicates that phone visits, nurse visits, and doctor visits were significant predictors at $p < 0.0001$, while self control and educational patient meetings were not so. Table 14.4 tests, whether the outliers are significantly more distant from the ARIMA model than could happen by chance. All of the p-values were very significant with $p < 0.001$ and < 0.0001.

5   Discussion

The current chapter shows that ARIMA is appropriate for assessing trends, seasonality, and change points in a time series. The current chapter shows that even with relatively small data a lot of information is obtained. Of course, there are additional ways for assessing each of these effects. For example, trends can be assessed by trend tests like chi-square and linear regression trend tests, seasonality

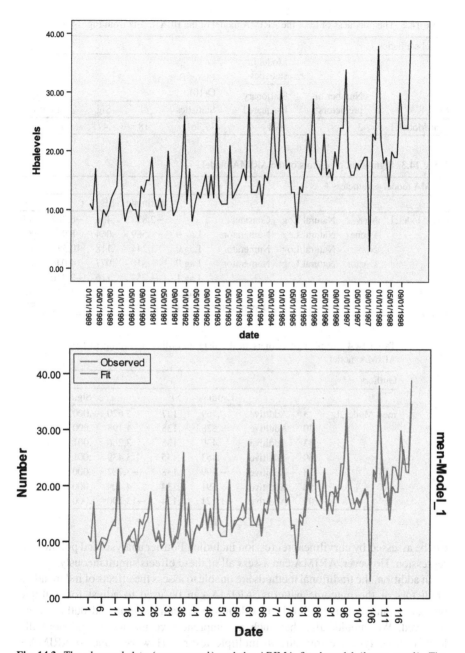

Fig. 14.3 The observed data (*upper graph*) and the ARIMA fitted model (*lower graph*). The fitted model very well matches the observed data. There are (1) numerous peaks, they are (2) approximately equally sized, and (3) there is an upward trend. Finally (4), there are several peaks that are not part of the seasonal pattern, may be due to outliers. Outliers are addressed using the Expert Modeler

Table 14.2 The goodness of fit of the ARIMA model of the HbA1c data from Fig. 14.3

Model statitstics

Model	Number of predictors	Model fit statistics	Ljung-Box Q(18)			Number of outliers
		Stationary R-squared	Statistics	df	Sig.	
men-Model_1	3	.898	17.761	18	.471	7

Table 14.3 Significant predictors of the ARIMA model

ARIMA model parameters

					Estimate	SE	t	Sig.
men-Model_1	Men	Natural Log	Constant		−2.828	.456	−6.207	.000
	Phone	Natural Log	Numerator	Lag 0	.569	.064	8.909	.000
	Nurse	Natural Log	Numerator	Lag 0	1.244	.118	10.585	.000
	Doctor	Natural Log	Numerator	Lag 0	.310	.077	4.046	.000
				Lag 1	−.257	.116	−2.210	.029
				Lag 2	−.196	.121	−1.616	.109
			Denominator	Lag 1	.190	.304	.623	.535

Table 14.4 Number of outliers in the data according according to the ARIMA model

Outliers

			Estimate	SE	t	Sig.
men-Model_1	3	Additive	.769	.137	5.620	.000
	30	Additive	.578	.138	4.198	.000
	53	Additive	.439	.135	3.266	.001
	69	Additive	.463	.135	3.439	.001
	78	Additive	−.799	.138	−5.782	.000
	88	Additive	.591	.134	4.409	.000
	105	Additive	−1.771	.134	−13.190	.000

can be assessed by curvilinear regression including Fourier analysis and polynomial regression. However, ARIMA can assess all of these effects simultaneously.

In addition, the traditional methods are unable to assess the effects of independent predictors of the outcome patterns. ARIMA can be used to adjust for multiple predictors, and is, thus, helpful to explain *why* certain patterns and outliers were observed. We should add, that such a comprehensive method would normally lead to large type I errors due to multiple testing. However, due to ARIMA's multivariate methodology, this is prevented and the very sensitivity of testing is even improved [4].

Autoregressive integrated moving average (ARIMA) modeling can include the effects of multiple predictors on the outcome patterns. But it is different from multiple linear regression, because, instead of a single continuous outcome variable, multiple outcome variables along a time line are used. It is also different from

Cox regression, because Cox regression assesses time to a single event, rather than specific patterns in a time line.

ARIMA can also be applied to analyze time series in individual patients. We should emphasize that n = 1 studies are not scientific trials, because they are not meant to improve the level of scientific knowledge of the community, but rather to find the best possible treatment for individual patients. ARIMA is a sensitive way to assess up/downward trends and change points along the time line, and this may be so, not only for groups, but also for individual patients. It has been successfully applied as a methodology for the latter purpose [9, 10].

The current chapter reports summary statistics of a population of patients without taking the individual values into account. Summary statistics are often used to describe populations and they are unbiased estimators, but, because they do not include a measure of spread, information will be lost. Indeed, no conclusions can be drawn about the individual members of a population. However, conclusions about the population at large are no problem, and can be relevant, like the effects of seasons and treatment changes on populations with chronic diseases.

6 Conclusions

1. ARIMA modeling is appropriate for assessing trends, seasonality, and change points in a time series.
2. It can assess all of these effects simultaneously.
3. It can also be used to adjust for multiple predictors, and is, thus helpful to explain *why* certain patterns and outliers were observed.
4. It is a multivariate methodology, preventing thereby the type I error risk of multiple testing.
5. It is different from multiple linear regression, because, instead of a single continuous outcome variable, multiple outcome variables along a time line are used.
6. It is also different from Cox regression, because Cox regression assesses time to a single event, rather than pattern characteristics along a time line.
7. It can also be applied to analyze time series in individual patients.
8. It can be used to analyze summary statistics of large populations, like the effects of seasons and treatment changes on populations with chronic diseases.

Appendix

The data file of example. (Date = date of observation, HbA1 = mean HbA1c of diabetes population, nurse = mean number of diabetes nurse visits, doctor = mean number of doctor visits, phone = mean number of phone visits, self = mean number of self-controls, meeting = mean number of patient educational meetings)

Date	HbA1	Nurse	Doctor	Phone	Self	Meeting
01/01/1989	11.00	8.00	7.00	3	22	2
02/01/1989	10.00	8.00	9.00	3	27	2
03/01/1989	17.00	8.00	7.00	2	30	3
04/01/1989	7.00	8.00	9.00	2	29	2
05/01/1989	7.00	9.00	7.00	2	23	2
06/01/1989	10.00	8.00	9.00	3	27	2
07/01/1989	9.00	8.00	8.00	3	27	2
08/01/1989	10.00	8.00	7.00	3	30	2
09/01/1989	12.00	8.00	8.00	4	27	2
10/01/1989	13.00	9.00	11.00	3	32	2
11/01/1989	14.00	9.00	7.00	3	29	2
12/01/1989	23.00	10.00	11.00	5	39	3
01/01/1990	12.00	8.00	7.00	4	23	2
02/01/1990	8.00	8.00	6.00	2	25	3
03/01/1990	10.00	9.00	6.00	3	30	3
04/01/1990	11.00	8.00	8.00	4	21	2
05/01/1990	10.00	9.00	9.00	3	27	2
06/01/1990	10.00	9.00	8.00	3	26	3
07/01/1990	8.00	9.00	6.00	2	36	3
08/01/1990	14.00	9.00	7.00	3	25	2
09/01/1990	13.00	9.00	9.00	3	27	2
10/01/1990	15.00	9.00	11.00	3	31	2
11/01/1990	15.00	9.00	9.00	3	26	3
12/01/1990	19.00	11.00	9.00	3	26	3
01/01/1991	12.00	9.00	7.00	3	26	3
02/01/1991	10.00	8.00	7.00	3	25	3
03/01/1991	12.00	9.00	7.00	3	29	3
04/01/1991	10.00	9.00	8.00	2	29	3
05/01/1991	10.00	9.00	9.00	2	34	2
06/01/1991	20.00	10.00	8.00	2	27	3
07/01/1991	13.00	9.00	6.00	3	29	2
08/01/1991	9.00	9.00	9.00	2	29	3
09/01/1991	10.00	10.00	9.00	2	25	2
10/01/1991	12.00	9.00	9.00	2	27	2
11/01/1991	16.00	9.00	8.00	3	28	2
12/01/1991	26.00	12.00	8.00	4	34	3
01/01/1992	11.00	9.00	9.00	3	23	3
02/01/1992	17.00	9.00	5.00	3	33	3
03/01/1992	8.00	9.00	7.00	2	29	3
04/01/1992	11.00	9.00	11.00	3	28	3
05/01/1992	13.00	9.00	5.00	4	29	3
06/01/1992	12.00	10.00	7.00	3	29	4
07/01/1992	14.00	9.00	9.00	4	28	4
08/01/1992	16.00	10.00	9.00	4	27	4
09/01/1992	12.00	9.00	8.00	3	34	3

(continued)

(continued)

Date	HbA1	Nurse	Doctor	Phone	Self	Meeting
10/01/1992	16.00	10.00	9.00	3	31	4
11/01/1992	12.00	9.00	8.00	3	31	3
12/01/1992	26.00	13.00	9.00	4	28	5
01/01/1993	12.00	9.00	8.00	4	27	4
02/01/1993	11.00	10.00	6.00	3	28	4
03/01/1993	11.00	10.00	8.00	3	29	4
04/01/1993	11.00	10.00	8.00	3	30	4
05/01/1993	21.00	10.00	8.00	3	31	4
06/01/1993	12.00	9.00	7.00	3	27	4
07/01/1993	13.00	10.00	9.00	3	34	4
08/01/1993	14.00	10.00	8.00	3	29	3
09/01/1993	15.00	10.00	8.00	3	29	4
10/01/1993	17.00	10.00	8.00	3	32	4
11/01/1993	15.00	10.00	9.00	3	32	4
12/01/1993	28.00	13.00	9.00	4	34	5
01/01/1994	13.00	10.00	7.00	4	27	3
02/01/1994	13.00	10.00	8.00	3	28	3
03/01/1994	13.00	10.00	7.00	3	26	3
04/01/1994	15.00	10.00	7.00	4	29	4
05/01/1994	11.00	10.00	7.00	3	32	4
06/01/1994	16.00	11.00	8.00	4	28	3
07/01/1994	15.00	10.00	10.00	4	27	3
08/01/1994	18.00	11.00	7.00	4	30	3
09/01/1994	27.00	10.00	9.00	4	32	4
10/01/1994	19.00	10.00	9.00	4	23	4
11/01/1994	17.00	10.00	8.00	3	32	4
12/01/1994	31.00	13.00	10.00	5	31	5
01/01/1995	24.00	10.00	8.00	5	31	4
02/01/1995	17.00	10.00	8.00	5	30	4
03/01/1995	18.00	11.00	5.00	5	27	4
04/01/1995	13.00	10.00	9.00	3	25	5
05/01/1995	13.00	10.00	8.00	3	29	6
06/01/1995	7.00	11.00	8.00	3	29	5
07/01/1995	14.00	11.00	7.00	3	29	4
08/01/1995	13.00	10.00	10.00	3	25	4
09/01/1995	17.00	10.00	5.00	4	29	3
10/01/1995	23.00	11.00	11.00	5	24	4
11/01/1995	19.00	11.00	9.00	4	29	4
12/01/1995	29.00	15.00	10.00	4	40	5
01/01/1996	18.00	10.00	8.00	5	31	4
02/01/1996	17.00	12.00	7.00	4	31	4
03/01/1996	16.00	11.00	7.00	4	31	4
04/01/1996	23.00	10.00	8.00	3	30	4
05/01/1996	16.00	11.00	8.00	3	27	4

(continued)

(continued)

Date	HbA1	Nurse	Doctor	Phone	Self	Meeting
06/01/1996	17.00	11.00	9.00	4	27	4
07/01/1996	15.00	11.00	8.00	4	31	3
08/01/1996	20.00	11.00	7.00	4	23	4
09/01/1996	17.00	11.00	10.00	4	29	5
10/01/1996	24.00	11.00	9.00	4	29	5
11/01/1996	24.00	11.00	9.00	5	24	4
12/01/1996	34.00	14.00	9.00	5	35	5
01/01/1997	18.00	12.00	8.00	4	24	4
02/01/1997	16.00	11.00	8.00	4	27	4
03/01/1997	16.00	11.00	6.00	4	29	3
04/01/1997	18.00	11.00	9.00	4	32	4
05/01/1997	17.00	11.00	6.00	4	32	5
06/01/1997	18.00	11.00	10.00	4	29	4
07/01/1997	19.00	12.00	7.00	4	26	5
08/01/1997	19.00	11.00	7.00	4	29	5
09/01/1997	3.00	11.00	9.00	4	30	5
10/01/1997	23.00	12.00	11.00	4	26	5
11/01/1997	22.00	11.00	8.00	4	21	5
12/01/1997	38.00	14.00	9.00	6	31	6
01/01/1998	22.00	12.00	9.00	5	24	5
02/01/1998	18.00	12.00	6.00	4	32	5
03/01/1998	19.00	12.00	9.00	5	27	5
04/01/1998	14.00	12.00	7.00	3	28	6
05/01/1998	20.00	12.00	9.00	5	18	4
06/01/1998	19.00	12.00	8.00	5	29	5
07/01/1998	19.00	11.00	8.00	5	28	4
08/01/1998	30.00	12.00	9.00	4	27	5
09/01/1998	24.00	13.00	8.00	5	30	5
10/01/1998	24.00	12.00	10.00	4	28	6
11/01/1998	24.00	11.00	8.00	5	26	5
12/01/1998	39.00	15.00	10.00	5	37	7

References

1. Yates F (1952) George Udney Yule. Obituary notices of the Royal Society of Statistics, vol 8, p 308
2. Box G, Jenkins G, Reinsel G (1994) Time series analysis: forecasting and control, 3rd edn. Prentice-Hall, Englewood Cliffs
3. SPSS statistical software. www.spss.com. 1 Dec 2012
4. Strom Anonymous. Autoregressive integrated moving average. www.wikipedia.org/wiki/Autoregressive_integrated_moving_average. 12 Dec 2012
5. Swedish Foundation for Strategic Research (2013) Predictive models and biomarkers for new drugs, 2010–2013. www.stratresearch.se. 6 Dec 2012
6. Branham M, Ross E, Govender T (2012) Predictive models for maximum recommended therapeutic dosages of drugs. Comput Math Methods Med. doi:10.1155/2012/469769

7. Linden A, Adams J (2003) Using an empirical method for setting expectations in quality improvement initiatives. Oral presentation, Dis Manage Assoc America, Annual meeting

8. Chen A, Sin L (2005) Using ARIMA models to predict and monitor the number of beds occupied during a SARS outbreak in Singapore. BMC Health Serv Res 36:36–40

9. Marshall N, Holzapple P. Melodic intonations therapy, variations on a theme. www.aphasiology.pitt.edu/archive/PDF, 12 Dec 2012

10. K, Boutsen F (2009) ARIMA procedures in single subject research: an evolution of best practices in statistical methodology. Aphasiol Archive; 39th clinical aphasiology conference. Keystone, CO, USA, 26–30 May 2009. ID code 2091

11. Sato C. Disease management with ARIMA model in time series. www.academia.edu/2068369/Disease_Management, 12 Dec 2012

12. Tseng C, Brimacombe M, Xie M, Rajan M, Wang H, Kolassa J, Crystal S, Chen T, Pogach L, Safford M (2005) Seasonal patterns in monthly hemoglobin A1c. Am J Epidemiol 161:565–574

Chapter 15
Support Vector Machines

1 Summary

1.1 Background

A support vector machine (SVM) is the name for a specific supervised learning model, and is used for pattern recognition. It can be applied both for classification and for regression analysis.

1.2 Objective and Methods

Using the 2,400 patient Dutch national register of individuals with familial hypercholesterolemia as example, we will describe the background of SVMs, and, finally, report the result of the SVM in successfully classifying cases and controls. We will also compare these results to what can be achieved with logistic regression and discriminant analysis.

1.3 Results

The SVM classification results of the training set provided a specificity of 0.97 and a sensitivity of 0.37. In the test set specificity was 0.93 and sensitivity 0.28. With increasing C-value (penalty-parameter) specificity decreased and sensitivity increased. Standard logistic regression analysis and linear discriminant analysis both had a specificity of 0.91 and sensitivity of 0.31.

T.J. Cleophas and A.H. Zwinderman, *Machine Learning in Medicine:*
Part Two, DOI 10.1007/978-94-007-6886-4_15,
© Springer Science+Business Media Dordrecht 2013

1.4   Conclusions

1. SVMs offer much flexibility especially with respect to modeling non linear relations between patient-characteristics and outcome variables.
2. They are also flexible in handling interactions between patient-characteristics. In this sense SVMs have much in common with neural networks and other machine learning techniques.
3. In general, SVMs seem to be somewhat more stable than others, but heuristic studies to indicate whén SVM performs better than other machine learning tools are missing.
4. SVMs have similar advantages over more traditional statistical models, including the possibility to process imperfect data and complex non linear data (and, in the example given, providing a better sensitivity/specificity for modeling dependent variables than did traditional logistic regression, respectively 97 % and 37 % versus 91 % and 31 %), but, here too, it is difficult to point out, whén SVM performs better than these traditional tools do.

2   Introduction

A support vector machine (SVM) is the name for a specific supervised learning model and its associated computerized learning algorithm [1]. It is used for pattern recognition and can be applied both for classification and for regression analysis. Most often it is used for classification and it is then an alternative for discriminant analysis in all its variants, (multinomial) logistic regression, classification trees and neural network analysis. SVMs are used in many different application areas ranging from robotics, to weather prediction, language analysis and text mining, to image and signal analysis, and to DNA comparison, and protein modeling. In clinical medical research SVMs have been used to develop prediction models both for disease diagnosis and prognosis given a specific diagnosis. Here we will discuss the most basic SVM and apply it to the problem of classifying patients with cardiovascular events (cases) and patients without such events (controls) using a collection of patient characteristics. We will first describe the data of this example, then describe the background of SVMs and finally report the result of the SVM in successfully classifying the cases and the controls. We will also compare these results to what can be achieved with logistic regression and discriminant analysis.

3   Example Data

The data that we use to illustrate support vector machines comes from a cohort study of patients with familial hypercholesterolemia in the Netherlands [2]. The cohort consists of a random sample of 2,400 patients taken from the Dutch national

Table 15.1 Characteristics of patients with and without cardiovascular events

	Patients without events (n = 1,618)	Patients with events (n = 782)
Male gender: n (%)	693 (43 %)	487 (62 %)
BMI (kg/m²): mean (sd)	24.8 (3.6)	25.7 (3.3)
Smoking ever: n (%)	813 (77 %)	518 (88 %)
Systolic blood pressure (mmHg): mean (sd)	133 (18)	138 (22)
Diastolic blood pressure: (mmHg): mean (sd)	81 (10)	83 (11)
Hypertensive: n (%)	96 (6 %)	134 (17 %)
Total cholesterol (mmol/L): mean (sd)	9.5 (1.9)	9.7 (2.2)
HDL cholesterol (mmol/L): mean (sd)	1.24 (0.36)	1.14 (0.33)
Triglycerides (mmol/L): mean (sd)	1.69 (0.93)	2.06 (1.18)
Glucose (mmol/L): mean (sd)	4.96 (0.89)	5.33 (1.21)
Hba1c (%): mean (sd)	5.64 (1.11)	6.09 (1.41)
Diabetic: n (%)	51 (3 %)	87 (11 %)
Creatinine (μmol/L): median (IQR)	79 (19)	83 (21)
Apolipoprotein(a) (nmol/L): median (IQR)	150 (330)	230 (553)
Homocystein (μmol/L): median (IQR)	10.7 (4.2)	12.0 (5.)

IQR Interquartile range

register of individuals with familial hypercholesterolemia. Most patients carried a mutation of the LDL-receptor gene, but there were carriers of mutations in other genes as well and there were also individuals from families without any of the known mutations that cause life-long increased LDL-cholesterol levels. Patients were followed from birth and the event of interest here was the premature occurrence of any cardiovascular event, i.e., before age 50 years in males and before age 55 years in female patients. There were 782 patients with events (33 %) and patient characteristics that were considered as predictive of a cardiovascular event, are reported in Table 15.1. All characteristics differed significantly between cases and controls.

4 Methodological Background of Support Vector Machines

Basic aim of SVMs is to construct a hyperplane formed by the set of patient characteristics that separates the cases and controls as good as possible. For two dimensional data this hyperplane is the best fit separation line, for three dimensional data the best fit separation plane. For multidimensional data hyperplanes equally exist, although they are difficult to imagine. The aim of SVMs is similar to that of neural networks (and other techniques), but SVMs are usually better capable of finding the best fit hyperplane and are also more easily extended to patterns that are not linearly separable. This latter is done by transforming the data into "a new space" by Kernel functions, as non-parametric way to estimate random variables.

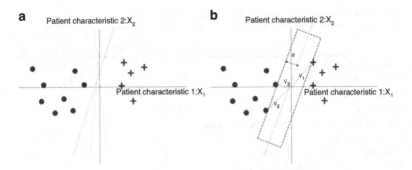

Fig. 15.1 Hyperplanes separating cases and controls

Consider the bivariate scatterplot of two patient-characteristics in Fig. 15.1a, where cases are denoted by crosses and controls by dots. A linear hyperplane is defined as $a*X_1 + b*X_2$, and "a" and "b" are weights associated with the two patient-characteristics. The object is to find "a", "b" and "c" such that $a*X_1 + b*X_2 \leq c$ for all controls and $a*X_1 + b*X_2 > c$ for cases. There are often many different solutions for "a", "b" and "c" and Fig. 15.1a shows three different hyperplanes that separate cases and controls equally well. The object of any classification technique is to find the optimal hyperplane, but the various techniques differ in which observations are used to define what is optimal. Regression models and neural networks apply all of the observations, but SVMs use only the "difficult" observations that are lying close to the optimal hyperplane (denoted as the "decision boundary"); these observations are called the support vectors. The rationale of this choice may be argued by the fact that the decision boundary will not change very much if the "easy" observations are removed from the dataset whereas the decision boundary will change dramatically if one or more of the difficult observations are removed. Therefore, the difficult observations or support vectors are the critical observations in the dataset. SVMs choose the hyperplane that maximizes the distance "d" from it to the difficult observations on either side (see Fig. 15.1b). The starting points of the support vectors denoted as v_1, v_2, and v_3 are in Fig. 15.1b. All three vector arrows meet in the origin. One line through the starting points of v_2 and v_3 is drawn, and one parallel line through the starting point of v_1. The best fit hyperplane line is midway between the two parallel lines, distance d (Fig. 15.1b). The distance from the hyperplane line to the origin is an important estimator in SVM statistics, and is expressed just like t-values in t-tests in a standardized way: if d = w (weight), and the distance from the hyperplane line to the origin = b, then the distance from the hyperplane line to the origin equals b/w (expressed in "w-units"). In order to extent this fairly simple procedure to more complex situations like multidimensional data and non linear data, a more general notation is preferred. It is given underneath.

Hyperplanes H are defined as $w'x_i + b \leq -1$ for controls and $w'x_i + b \geq 1$ for cases, where x_i is the vector of all patient-characteristics of patient i, w is a vector with weights and b is called a bias term and is comparable to an intercept of

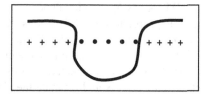

Fig. 15.2 A non-linear function might exist that is capable of linear separation, for instance the situation of the following sequence of cases and controls ranked on a single patient characteristic: $+ + + + \bullet\bullet\bullet\bullet + + + +$. Some cases have low values of the characteristic and other cases have high values, whereas controls have values in the middle of the distribution. There exists no linear hyperplane that separates cases and controls, but a simple quadratic function will do the job

regression models. The optimal set of weights is determined so that the distance "d" is maximized. It turns out that this is equivalent to minimizing the Euclidean norm $||w||^2$ subject to the condition that there are no observations in the margin: $w'x_i + b \le -1$ for controls and $w'x_i + b \ge 1$ for cases, or $y_i(w'x_i) \ge 1$ when $y_i = 1$ for cases and $y_i = -1$ for controls. This is a quadratic programming problem that can be solved by, for instance, the Lagrangian multiplier method.

This algorithm has been extended for situations where no hyperplane exists that can split the cases and controls. The so-called soft margin method chooses a hyperplane that separates the cases and controls as good as possible, but still maximizing the distance "d" to the nearest cleanly split observations. The method introduces slack-variables ξ_i that represent the degree of misclassification: $y_i(w'x_i) \ge 1 - \xi_i$. Instead of minimizing $||w||^2$ now $0.5||w||^2 + C*\Sigma_i \xi_i$ is minimized where the constant C is a penalty-parameter that regulates the amount of misclassification that is accepted.

In some cases where no linear separation exists, a non-linear function might exist that is capable of linear separation. Take for instance the situation of the following sequence of cases and controls ranked on a single patient-characteristic:

$$+ + + + \bullet\bullet\bullet\bullet + + + ++$$

Apparently, some cases have low values of the characteristic and other cases have high values, whereas controls have values in the middle of the distribution. There exists no linear hyperplane that separates cases and controls, but a simple quadratic function will easily do the job (Fig. 15.2).

This situation is not unlikely; take for instance a patient-characteristic such as body temperature: both when it is increased and when it is decreased, the risk of having a (infectious) disease might be large.

Transformations are, usually, considered through Kernel functions $K(x_i, x_j)$ of the inner products of the patient-characteristics of two observations x_i and x_j. The usual Kernel functions that are considered, are the polynomial Kernel function $(x_i'x_j + 1)^p$, the radial basis function $exp(-||x_i - x_j||^2/2\sigma^2)$ and the sigmoid function $tanh(\kappa\, x_i'x_j - \delta)$. The different transformations are characterized by parameters p, σ, and (κ, δ).

The effectiveness of a SVM for any application depends (next to that of the data) on the possibility to specify appropriate Kernel-functions, Kernel parameters, and soft margin C. Commonly, one starts with a Gaussian Kernel and then a grid search is performed for the parameters C and σ. Overfitting is a realistic danger for SVMs and it is crucial to do a form of double (and repeated) cross validation to select the optimal Kernels, parameters and C values and estimate the classification ability of the SVM.

Several software packages exist to develop a support vector machine. The free R system has an implementation in the package e1071. SPSS offers SVM models in the SPSS modeler package, SAS offers SVMs in their enterprise Miner package. It is also available in the standalone machine-learning program Weka (http:// weka.wikispaces.com/Primer). All offer SVM models for binary and multiclass classification as well as support vector regression analysis and SVM models for novelty detection.

5   Results of Fitting a Support Vector Machine to Our Example Data

We used the implementation of SVMs in the R-package e1071. We divided the total dataset in a training set of 1,600 patients and a test set of the remaining 800 patients. The SVM was developed with the data of the training set only; thus the optimal Kernel-function, Kernel-parameters and C-value were determined by cross validation of the training set. Classification results were then determined by applying the SVM to the data in the test set.

Radial basis Kernel functions yielded better results than linear, polynomial or sigmoid Kernels. Optimal Kernel-parameter σ^2 was 7.32 which was very close to the default choice in the R-package and the optimal C-value was 2. At the optimal value the classification results of the data in training set was a specificity of 0.97 and a sensitivity of 0.37. In the test set specificity was 0.93 and sensitivity 0.28. With increasing C-values specificity decreased and sensitivity increased. Standard logistic regression analysis and linear discriminant analysis both had specificity 0.91 and sensitivity 0.31.

6   Conclusions

Support vector machines are a valuable addition to the toolbox of any data-analyst.

1. It offers much flexibility, especially, with respect to modeling non linear relations between patient-characteristics and outcome variables.
2. It is also flexible in handling interactions between patient-characteristics. In this sense SVMs have much in common with neural networks and other machine learning techniques.

3. In general SVMs seem to be somewhat more stable than others, but heuristic studies to indicate whén SVMs perform better than other machine learning tools are missing.
4. SVMs have similar advantages over more traditional statistical models, including the possibility to process imperfect data and complex non linear data, but, here too, it is difficult to point out whén SVMs perform better than these traditional tools.

References

1. Cortes C, Vapnik V (1995) Support-vector networks. Mach Learn 20:273–297
2. Jansen AC, van Aalst-Cohen ES, Tanck MW, Trip MD, Lansberg PJ, Liem AH, van Lennep HW, Sijbrands EJ, Kastelein JJ (2004) The contribution of classical risk factors to cardiovascular disease in familial hypercholesterolaemia: data in 2400 patients. Intern Med 256:482–490

Chapter 16
Bayesian Networks

1 Summary

1.1 Background

A Bayesian network (BN) is a tool to describe and analyze multivariate distributions. In biomedicine the main applications are in expert systems, in bioinformatics applications in genetics, and in identifying gene-regulatory networks, with a particular eye towards causal relationships.

1.2 Objective

To assess whether a best fit Bayesian network can provide additional information about possible causal subgroup effects in clinical trials.

1.3 Methods

A randomized clinical 884 patient trial that evaluated efficacy of pravastatin to reduce cardiovascular events and to reduce the decrease of the diameter of coronary vessels was applied as example. Matlab Bayes Net toolbox was used to assess the role of age, smoking, and hypertension.

1.4 Results

The best fitting Bayesian network was much more complicated than expected involving dependencies between age and smoking, between smoking and baseline

T.J. Cleophas and A.H. Zwinderman, *Machine Learning in Medicine:*
Part Two, DOI 10.1007/978-94-007-6886-4_16,
© Springer Science+Business Media Dordrecht 2013

hypertension, LDL- and HDL-cholesterol and of all of these on LDL- and HDL-cholesterol change and change of vessel diameter. Coronary event appeared, in contrast, to be dependent only on randomized treatment and not on LDL- or HDL-decrease.

1.5 Conclusions

1. The graphical display of Bayesian networks has been adopted very widely, for instance, by biomolecular scientists to describe pathogenic, metabolic, and other causal pathways.
2. The graphical tools have proved to be appealing in every applied field, but are somewhat problematic in biomedicine, because confounding and interacting effects are often involved.
3. Nonetheless, Bayesian networks are very efficient to describe multivariate distributions. The structure makes inferences from Bayesian networks robust, reduces variances of estimated parameters, and is also robust against overfitting.

2 Introduction

A Bayesian network (BN) is a tool to describe and analyze multivariate distributions [1]. The tool is member of the family of probabilistic graphical models. Despite its association, Bayesian networks do not necessarily use Bayesian statistical methods for data-analysis, the name refers rather to the way conditional and marginal probability distributions are related to each other. Bayesian networks are sometimes used in conjunction with Bayesian statistical methods, especially when prior knowledge is analyzed together with data, but this is not unique for BNs. Graphical models in general, and Bayesian networks too, have been proposed especially to deal with complex data (−analysis), and with an eye towards causal interpretations. This is achieved by combining graph theory, probability theory, statistics and computer science. Bayesian networks have been used in many different fields, for instance, in the Microsoft Windows system and the NASA mission control. In biomedicine the main applications seems to be in expert systems, in bioinformatics applications in genetics, and in identifying gene-regulatory networks.

3 Example Data

The data that we will use to illustrate Bayesian networks, comes from a randomized clinical trial that evaluated efficacy of pravastatin to reduce cardiovascular events, and to reduce the decrease of the diameter of coronary vessels [2]. The trial consisted

Table 16.1 Characteristics of patients in two treatment groups (*Pravast* pravastatin)

	Placebo (n = 434)	Pravast (n = 450)	P – value
Coronary Event: n (%)	79 (18 %)	48 (11 %)	0.001
Decrease of the mean diameter of the coronary vessels (mm): mean (SD)	0.10 (0.21)	0.06 (0.19)	0.014
LDL-cholesterol decrease (mmol/L): mean (SD)	−0.04 (0.59)	1.23 (0.68)	< 0.001
HDL cholesterol increase (mmol/L): mean (SD)	0.03 (0.13)	0.10 (0.16)	< 0.001
Age (years): mean (SD)	56 (8)	57 (8)	0.26
LDL-cholesterol level at baseline (mmol/L): mean (SD)	4.31 (0.78)	4.30 (0.78)	0.75
HDL-cholesterol level at baseline (mmol/L): mean (SD)	0.93 (0.23)	0.93 (0.23)	0.72
Baseline mean diameter coronary vessels (mm): mean (SD)	2.82 (0.48)	2.80 (0.46)	0.46
Smoking ever: n (%)	376 (87 %)	402 (89 %)	0.22
Current hypertension: n (%)	134 (31 %)	112 (25 %)	0.06

of a random sample of 884 men with cardiovascular heart disease and normal to moderately enhanced LDL-cholesterol levels. Patients were randomized between 20 mg pravastatin DD or placebo for 2 years. Outcome variables were the change in mean diameter of the coronary segments measured at baseline and after 2 years with coronary angiography, and occurrence of coronary events during follow-up (death, myocardial infarction, stroke, coronary intervention). Four hundred fifty patients were randomized to pravastatin treatment and 434 to placebo. At baseline there were no significant or substantial differences between the two groups with respect to age, baseline LDL- and HDL-cholesterol, smoking history, and current hypertension. Also the average diameter of the coronary vessels did not differ between the groups at baseline. After 2 years of follow-up the number of patients with a coronary events was significantly lower in the patients who were treated with pravastatin, the decrease of the diameter of the coronary vessels was also significantly lower, and the change of LDL- and HDL-cholesterol levels was also significantly larger in the statin treated patients (Table 16.1).

Statins were supposed to work mainly through improving lipid levels, which would, in its turn, decrease the atherosclerotic process of lipid deposition in the wall of the coronary vessels. This process would, then, lead to lower risk of coronary events. This description forms a typical causal hypothesis that can be analyzed with graphical models.

4 Methodological Background of Bayesian Networks

Bayesian networks are probabilistic graphical models and use graphical structures to represent knowledge. In particular *nodes* and *edges* are used, where a node represents a random variable and an edge between two nodes represents a probabilistic

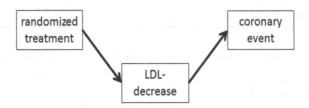

Fig. 16.1 A simple directed acyclic graph of the causal effect of the statin treatment-variable on LDL-decrease and the occurrence of coronary events during the follow-up of a clinical trial of patients with coronary heart disease

dependency between two variables. If edges are undirected the graphical models are usually called Markov networks of Markov random fields, but in Bayesian networks edges, usually, have direction and the graph is, then, called a directed acyclic graph (DAG). With such a DAG the multivariate distribution of the variables in the DAG can be represented efficiently, and the DAG provides also an easy way to estimate the distribution.

The directed edge from variable X_i to variable X_j represents the statistical dependence of X_j and X_i, but slightly stronger, X_i is defined to influence X_j or be X_j's parent (X_j is defined to be X_i's child). In more general terms, X_j is X_i's descendant and X_i is X_j's ancestor. The DAG is acyclic and that guarantees that a variable cannot be its own descendant or ancestor. A directed edge from X_i to X_j is often understood to represent a *causal* relationship between the two variables X_i and X_j.

For our example data, we can consider the graph in Fig. 16.1 of the outcome variable "coronary event: yes/no", the randomized treatment variable "pravastatin: yes/no", and the "LDL-cholesterol decrease". The directed edge between "random-ized treatment" and "LDL-decrease" points to our expectation that the choice of treatment will influence how much LDL-cholesterol will decrease. The directed edge between "LDL decrease" and "coronary event" points to our expectation that the amount of LDL-decrease, in its turn, will determine the risk of a coronary event. Thus, "randomized treatment" is an ancestor of both "LDL-decrease", and "coronary event", and "LDL-decrease" is also an ancestor of "coronary event".

BNs have rather simple conditional dependence and independence statements. The (directed) edge between "randomized treatment" and "LDL-decrease" signifies a direct dependence, namely that the distribution of the latter depends on the specific value of the former. But far more general, one may say that each variable is independent of its nondescendants in the graph given the state of its parents. Thus, "coronary event" is independent of "randomized treatment" given that the amount of "LDL-decrease" is known. Note that this particular DAG corresponds to the *causal* hypothesis that statin-treatment works only through lowering LDL-cholesterol level, and, thus, excludes a pleiotropic effect of statins.

In addition to the DAG, the network is represented by a set of conditional prob-ability distributions that together describe the multivariate distribution $L(X,Y,Z)$

where X = "randomized treatment", Y = "LDL-decrease" and Z = "coronary event". For the above DAG we can describe L(X,Y,Z) as the product L(X)*L(Y|X)*L(Z|Y), where L(Y|X) means the likelihood of Y given X. Given that the randomized treatment is determined by chance, L(X) would typically be described by a Bernoulli distribution with probability 0.5 (i.e. "throwing a coin"). LDL-decrease is a normal distributed quantity and L(Y|X) would therefore be a conditional distribution usually described with ordinary linear regression and because coronary event is a binary variable L(Z|Y) would typically be described by logistic regression. Leaving out the directed edge from "randomized treatment" to "coronary event" means that L(X,Y,Z) is less complex than a saturated model, meaning that one parameter less needs to be estimated from the data. This is not very imposing, but BNs of many variables can benefit greatly from assuming structure in the sense that inference is computationally much cheaper, and results can be far more robust with far less variance. The complexity of the multivariate distribution modeled with a BN is quantified by the so-called *d-separation* statistic.

Inference in a Bayesian network is done by marginalization, meaning that irrelevant variables are integrated or summed out. If the risk of a coronary event must be calculated for patients treated with pravastatin, this is calculated by $L(Z|X = statin) = \int L(Z|Y = y)\ L(Y = y|X = statin)\ dy$. Basically, the likelihood of a coronary event for all possible LDL-decreases Y = y are considered (i.e. L(Z|Y = y)) and these likelihoods are averaged but weighted with the likelihood that such a LDL-decrease Y = y is observed under statin treatment (i.e. L(Y = y|X = statin)). For a particular variable in a general Bayesian network this marginalization can be done through either its parents or its children, and the former is called *predictive support* or *top-down reasoning* while the latter is called *diagnostic support* or *bottom-up reasoning*. Which strategy is chosen, is determined for opportunistic reasons, but if the Bayesian network is large, exact inference may be very hard involving multiple integrals or summations. Popular exact algorithms are message-passing, cycle-cutset and variable-elimination. Approximate algorithms are useful for large Bayesian networks and are mostly based on Monte Carlo sampling such as the Markov Chain Monte Carlo (MCMC) methods.

Learning a new BN from data presents several difficulties: the Bayesian network structure may be known or unknown, the shapes of the conditional distributions $L(X_j|X_i)$ and their parameters may be known or unknown, and the variables in the Bayesian network may be observed or only partially observed. Given a particular Bayesian network structure and appropriate data, the best parameters describing the multivariate distribution are found by maximization of the log-likelihood of the data. This is fully comparable to estimating any statistical model. For the Bayesian network in Fig. 16.1 this would entail estimating the parameters of the linear regression model of LDL-decrease on randomized treatment and of the logistic regression model of coronary event on LDL-decrease. If the Bayesian network contains nodes for which no data is available, then the unobserved nodes must be partialed out. This can be done using MCMC methods or with expectation-maximization algorithms in less complex cases.

If the Bayesian network structure is unknown, the problem is, unfortunately, much harder, because the number of different DAGs with N variables is superexponential in N. In practice one then, usually, starts with a reasonably simple DAG (a naive Bayesian network for instance), and, then, adds those edges to the DAG that minimize some goodness of fit criterion such as the Akaike's or Bayesian Information Criterion (AIC/BIC). This is the approach we used for our example data.

When using Matlab Bayes Net toolbox [code.google.com/p/bnt/], the following syntax commands from Matlab prompt should be given.

1. For model selection:
   $$P(D|G) = \int_{\theta} p(D|G), \theta$$
2. For finding the best model:
   $$\sum_{k=0}^{n} \binom{n}{k} = 2^n$$
3. For computing BIC values:
   $$\log\Pr(D|G) \approx \log\Pr(D|G).\hat{\Theta}\_G) - \frac{\log N}{2}/\#G$$

5   Results of Fitting a Bayesian Network to Our Example Data

For our data we hypothesized that the randomized treatment only affected the risk of coronary events through lowering LDL-cholesterol. We assumed that this in its turn would reduce the decrease of the diameters of the coronary vessels. We hypothesized, in addition, that smoking affects the diameters of the coronary vessels, and that hypertension has a direct effect on the risk of coronary events. The DAG is illustrated in Fig. 16.2.

The DAG has three parents, namely "age", "smoking", and "randomized treatment". The structure also hypothesizes, that the significant association between randomized treatment and coronary events (see Table 16.1) will vanish by conditioning on change of vessel diameter. Similarly, the associations between age and smoking and coronary events will vanish after conditioning on hypertension and change of vessel diameter. The AIC values of the hypothesized DAG in Fig. 16.2 and five adaptations are reported in Table 16.2. The optimal AIC value was found for a model in which the conditional distribution of vessel-diameter-change depended directly on smoking and randomized treatment, and that the risk of a coronary event depended on the amount of LDL- and HDL- cholesterol change, and, in addition, on the randomized treatment. This latter result may be interpreted as the pleiotropic effect of pravastatin.

The best fitting DAG was, however, much more complicated, involving dependencies between age and smoking, between smoking and baseline hypertension, LDL- and HDL-cholesterol, and of all of these on LDL- and HDL-cholesterol change and change of vessel diameter. Coronary event appeared, in contrast, to be dependent only on randomized treatment and not on LDL- or HDL-decrease.

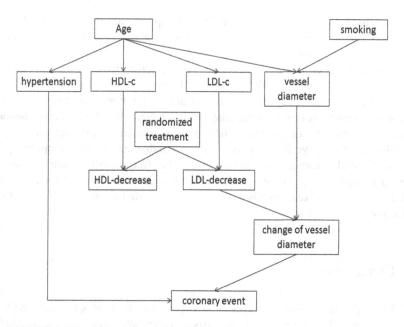

Fig. 16.2 Directed acyclic graph corresponding with the Bayesian network in which the effect of statin treatment has a direct effect on LDL-cholesterol and no pleiotropic effect on diameter of coronary vessels nor risk of coronary events

Table 16.2 Akaike information criterion values of the hypothesized Bayes network of Fig. 16.2 and five adaptations

Nr	Model	AIC
0	As illustrated in Fig. 16.2	13,147.26
1	Model 0 plus direct effects of LDL- and HDL-cholesterol decreases on events	13,145.15
2	Model 1 plus direct effect of randomized treatment on events	13,139.49
3	Model 1 plus direct effect of randomized treatment on change of diameter	13,143.06
4	Model 2 plus direct effect of randomized treatment on change of diameter	13,135.29
5	Model 3 plus direct effects of smoking on events and change of diameter	13,136.83

6 Discussion

Bayesian networks have attained much enthusiasm in many applied research fields. The graphical interface has proved to be very helpful both in summarizing relationships between a large set of variables and for hypothesizing about causality. The graphical display has been adopted very widely, for instance, by biomolecular scientists to describe pathogenic and metabolic pathways. But the graphical tools have proved to be appealing in every applied field. The causal interpretation of BNs is (in contrast) somewhat problematic in biomedicine. Causality is difficult in biomedical research, because, often, confounding effects can not be ruled out in observational data, and it is very difficult to specify the influence of selection processes for any given sample of patients. It is also difficult to do controlled experiments with human subjects.

Aside from interpretations, BNs are very efficient to describe multivariate distributions. The structure makes inference of BNs often robust and it also reduces variance of estimated parameters. Thus BNs are often robust against overfitting. In case a new network is learned from a dataset, it is nevertheless highly recommended to perform some form of cross validation to assess reliability of the network.

Software for Bayesian networks are available in many computer programs. Several packages are available in the freeware/shareware R system [www.r-project.org, package: deal], several algorithms are offered in the weka package [weka.sourceforce.net] and the Matlab Bayes Net toolbox [code.google.com/p/bnt/].

7 Conclusions

1. The graphical display of Bayesian networks has been adopted very widely, for instance, by biomolecular scientists to describe pathogenic and metabolic pathways.
2. The graphical tools have proved to be appealing in every applied field. The causal interpretation of BNs is (in contrast) somewhat problematic in biomedicine.
3. Causality is difficult in biomedical research, because, often, confounding effects can not be ruled out in observational data, and it is also difficult to do controlled experiments with human subjects.
4. Bayesian networks are very efficient to describe multivariate distributions. The structure makes inferences from Bayesian networks robust, reduces variances of estimated parameters, and is also robust against overfitting.

References

1. Ben-Gal I (2007) Bayesian networks. In: Ruggeri F, Faltin F, Kenett R (eds) Encyclopedia of statistics in quality & reliability. Wiley, Hoboken, NJ, USA
2. Jukema JW et al (1995) Effects of lipid lowering by pravastatin on progression and regression of coronary artery disease in symptomatic men with normal to moderately elevated serum cholesterol levels. The Regression Growth Evaluation Statin Study (REGRESS). Circulation 91:2528–2540

Chapter 17
Protein and DNA Sequence Mining

1 Summary

1.1 Background

In the past two or three decades the role of genetic determinants has increased enormously in biomedical research.

1.2 Objective

In this chapter we will review a number of these new techniques for the analysis of high throughput protein / DNA data, and for the analysis of gene-expression data.

1.3 Methods

Gene expression analysis is explained with help of microarrays and hierarchical cluster analysis. Sequence similarity searching is explained with the help of the Basic Local Alignment Tool (BLAST) from the US National Center of Biotechnology Information.

1.4 Results

The normalized ratios and log-transformed ratios of the gene expression levels identified 14 % over-expression cancer genes in a single patient with gastric cancer.

T.J. Cleophas and A.H. Zwinderman, *Machine Learning in Medicine:*
Part Two, DOI 10.1007/978-94-007-6886-4_17,
© Springer Science+Business Media Dordrecht 2013

Hierarchical cluster analysis of 18 patients with gastric cancer identified two clusters of patients. BLAST similarity search was used to assess various query amino acid sequences, and detected clinically relevant similarities, e.g., with streptococcus and nocardia proteins and with gentamicin resistant pseudomonas proteins.

1.5   Conclusions

1. With the help of gene expression analyses thousands of genetic variants and differentially expressed genes have been associated with clinical effects / functions, and, in addition, the proteins associated with these genes have been investigated to disentangle their roles in the biochemical and physiological pathways of the disease and treatments being studied.
2. Sequence similarity searching is a method that can be applied by almost anybody for finding similarities between his / her query sequences and the sequences known to be associated with different clinical effects.

2   Introduction

In the past two or three decades the role of genetic determinants has increased enormously in biomedical research. Of several monogenetic diseases the genetic foundation has been clarified almost completely (e.g., Huntington's disease), and of others the contribution of many genetic markers has been proved: for instance the brca 1 and 2 genes in breast cancer [1], and the mismatch gene mutations in colon carcinoma [2]. Simultaneously, the human genome project has been the catalyst for the development of several high throughput technologies that have made it possible to map and sequence complex genomes. These technologies are used, and will be used increasingly in clinical trials for many purposes but predominantly to identify genetic variants and differentially expressed genes that are associated with better or worse clinical efficacy in clinical trials. In addition, the proteins associated with these genes are being investigated to disentangle their roles in the biochemical and physiological pathways of the disease and the treatment that is being studied. Together these technologies are called (high throughput) genetics, genomics, and proteomics.

The technological advancements have made it possible to measure thousands of genes / proteins of a single patient simultaneously, and to evaluate the role of each gene/protein in differentiating between e.g., responders and non − responders to therapy. This has increased the statistical problem of multiple testing hugely, but also has stimulated research into statistical methods to deal with it. In addition methods have been developed to consider the role of clusters of genes. In this chapter we will review a number of these new techniques for the analysis of high throughput genetic data, and for the analysis of gene-expression data. We restrict the review to data that are typically sampled in clinical trials including unrelated individuals only. Familial data are extremely important to investigate genetic associations: their clustered structure requires dedicated statistical techniques but these fall outside the scope of this chapter.

3 Gene Expression Analysis

In the mid-1970s, molecular biologists developed molecular cloning and DNA sequencing. Automated DNA sequencing and the invention of the polymerase chain reaction (PCR) made it possible to sequence the entire human genome. This has lead to the development of microarrays, sometimes known as DNA-chip technology. Microarrays are ordered sets of DNA molecules of known sequence. Usually rectangular, they can consist of a few hundred to thousands of sets. Each individual feature goes on the array at a precisely defined location on the substrate, and thereafter, labeled cDNA from a test and a reference RNA sample are pooled and co-hybridized. Labeling can be done in several ways, but is usually done with different fluorescently labeled nucleotides (usually Cy5-dCTP for reference, and Cy3-dCTP for test RNA). After stimulation, the expression of these genes can be measured. This involves quantifying the test and reference signals of each fluorophore for each element on the array, traditionally by confocal laser scanning. The ratio of the test and reference signals is commonly used to indicate whether genes have differential expression. Many resources are available on the web concerning the production of microarrays, and about designing microarray experiments (e.g., 123genomics.homestead.com). A useful textbook is that of Jordan [3]. An example of a microarray is given in Fig. 17.1. This concerns the differential expression of about 500 genes in tumor tissue of a single patient with gastric tumor.

Each spot in this chip represents a different gene, and the ratio of the two fluorescent dyes indicates whether the genes are over-expressed (dark) or under-expressed (pale) in the tumor tissue with respect to normal tissue. The transformation of the image into gene expression numbers is not trivial: the spots have to be identified on the chip, their boundaries defined, the fluorescence intensity measured, and compared to the background intensity. Usually this 'image processing' is done automatically by the image analysis software, but sometimes laborious manual adjustments are necessary. One of the most popular systems for image analysis is ScanAlyze (http://rana.stanford.edu/software). After the image analysis, differential expression is measured by a so-called normalized ratio of the two fluorescence signals, normalized to several experimental factors. The normalized ratios of the array in Fig. 17.1 are given in Fig. 17.2. On the x-axis are given the 500 genes, and on the y-axis is given the normalized ratio of each gene. It is obvious that most genes have a ratio around unity, but three or four genes are highly over-expressed with ratios above two. It is typically assumed that ratios larger than 1.5 or 2.0 are indicative of a significant change in gene expression. These estimates are very crude,

Fig. 17.1 Example of microarray of different expression of about 500 genes in tumor tissue of a single patient

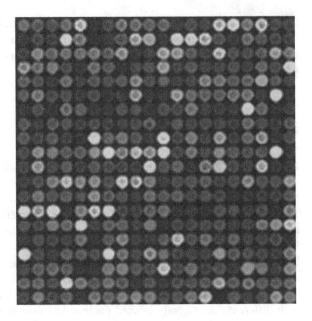

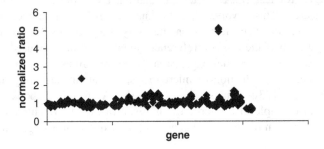

Fig. 17.2 Normalized ratios of the array from Fig. 17.1

however, because the reliability of ratios depends on the two absolute intensities. On statistical grounds, moreover, we would expect a number of genes to show differential expression purely by chance [4].

One way of circumventing the multiple testing problem here, is to use a mixture model [5]. Usually, it is assumed that the sample of ratios consists of subgroups of genes with normal, under-, and over-expression. In each subgroup, the ratios are mostly assumed to be normally distributed. When the sample is large enough, the percentage of normal, under-, and over-expressed genes, and associated mean ratios and standard deviations can be estimated from the data. This can be done with the logarithmically transformed ratios. The histogram of the log-transformed ratios in Fig. 17.2 is given in Fig. 17.3, together with the three estimated normal distributions. In this model the probability of each gene of being over- of under-expressed can be calculated using Bayes' theorem.

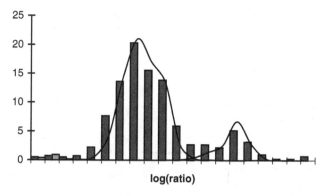

Fig. 17.3 The histogram of the log-transformed ratios from Fig. 17.2, calculated according to Bayes' theorem

Although under-expressed genes could not be identified in this case, over-expressed genes were clearly seen, represented by the second mode to the right. Actually it was estimated that 14 % of the genes showed over – expression, corresponding with ratios larger than 1.3. Above is illustrated how to look at the data of a single microarray. For the analysis of a set of microarrays several different approaches are used. Two distinctions can be used: supervised or unsupervised data analysis, and hypotheses – driven or data-mining. For supervised data analysis additional data must be available to which the expression data can be related. In clinical trials a major question is often how responders and non – responders can be distinguished. Relating such response data to expression data can be done using well known techniques such as discriminant – analysis, or logistic regression. Since there may be hundreds or thousands of expression variables, one must be careful in applying these techniques, and cross – validation is often extremely useful [6]. Unsupervised data analysis is usually done by cluster analysis or principal component analysis to find groups of co-regulated genes or related samples. These techniques are often applied without specific prior knowledge on which genes are involved in which case the analysis is a kind of data – mining. An example of a hypothesis driven analysis is to pick a potential interesting gene, and then find a group of similar or anti – correlated expression profiles. Cluster-analysis is the most popular method currently used as the first step in gene expression analysis. Several variants have been developed: hierarchical [7], and k-means [8] clustering, self-organizing maps [9], and gene-shaving [10], and there are many more. All aim at finding groups of genes with similar properties. These techniques can be viewed as a dimensionality reduction technique, since the many thousands of genes are reduced to a few groups of similarly behaving genes. Again many tools are available on the web, and a useful site to start searching is: www.microarray.org. We used Michael Eisen's package [7] to cluster the expression data of 18 patients with gastric cancer. The typical output of a hierarchical clustering analysis is given in Fig. 17.4. This is a dendrogram illustrating the similarities between patients, a similar graph can

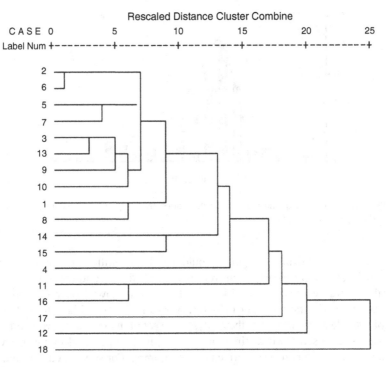

Fig. 17.4 The typical hierarchical clustering analysis of the expression data of 18 patients with gastric cancer

be obtained illustrating similarities between genes. In the present case one might conclude that patients 2, 6, 5, 7, 3, 13, 9, 10, 1 and 8 form a cluster, and patients 14, 15, 4, 11, 16, 17, 12, and 18 another cluster. But identifying more clusters may be meaningful too.

In a K-means cluster analysis the number of clusters must be specified a priori. When we specify two clusters, the same solution is found as above. The above results illustrate that many subjective decisions need to be made in a cluster analysis, and such analysis cannot be regarded as hypothesis-driven; the primary output of a cluster analysis are new hypotheses concerning differential expressions.

4 Sequence Similarity Searching

With the help of gene expression analyses thousands of genetic variants and differentially expressed genes have been associated with clinical effects / functions, and, in addition, the proteins associated with these genes have been investigated to disentangle their roles in the biochemical and physiological pathways of the disease and treatments being studied. Sequence similarity searching is a method that can be applied by almost anybody for finding similarities between his / her query sequences and the sequences known to be associated with different clinical effects. The latter have been included in database systems like the BLAST (Basic Local Alignment Search Tool) database system from the US National Center of Biotechnology Information (NCBI) [11], and the MOTIF data base system, a joint website from different European and American institutions [12], and they are available through the Internet for the benefit of individual researchers trying and finding a match for novel sequences from their own research. The information given by the data bases can be applied in clinical research in multiple ways. As an example, if one wishes to study the possibility of a cryptosporum infection, but does not have cryptosporum DNA for matching available, one could extract DNA from a specimen culture, use DNA polymerase to amplify the DNA, and thus obtain a DNA sequence at one's own laboratory. This DNA sequence could, then, be examined using, e.g., the interactive database of NCBI BLAST [11], and a statistically significant homology of the obtained DNA with known cryptosporum nucleotide sequences could be pursued.

How does the interactive system work? We will use protein sequences as example, but DNA sequences can be equally assessed. Both amino acid and nucleic acid sequences in living creatures are subject to deletions and mutations, and, therefore, proteins and DNA may be homologous (meaning that they stem from one and the same ancestor) in spite of being somewhat different. E.g., the amino acid sequence SIKLWPPSQTTR, respectively, contains the amino acids serine, isoleucine, lysine, isoleucine, trypthophan, proline, proline, glutamine, threonine, threonine, and arginine. If we assess this unknown sequence with BLAST, we will find several matches with known sequences including

Query sequence SIKLWPPSQTTR
database match 1 SIKLWPPSQ TR (found in GTP-ase activator protein isoform 1)
database match 2 SIKLWPPSQNTR (found in GTP-ase activator protein isoform 2)

In database match 1, a gap is observed: the first amino acid T (threonine) has been deleted. In database match 2 a change of amino acid is observed: the first amino acid T (threonine) is changed into N (asparagine), either due to mutation or to polymorphism.

Dayhoff (1978) [13] was the first to implicate the risk of changes in amino acid sequences, and constructed similarity scores for all possible pairs of amino acids, where similarity scores are defined as the estimated measure for the overall frequency that any two amino acids, like T versus T, or T versus N, etc., occur together in real life, based on numbers of sequences observed in the database applied. The similarity score of the amino acid "a" versus "b" are expressed by the equation (log = natural logarithm)

$$\text{Similarity score} = \log [\text{frequency(ab)}] / [\text{frequency(a).frequency(b)}],$$

and the risk of deletions (gaps) was also accounted. Amino acids that occur together more often than expected by chance have positive scores, those that occur less

Table 17.1 Part of the
BLOSUM 62 matrix of
similarity alignment scores
between individual amino
acids (*ala* alanine, *arg*
arginine, *asn* asparagine, *asp*
aspartic acid, *cyst* cysteine)

	ala	arg	asn	asp	cys
ala	4				
arg	−1	5			
asn	−2	0	6		
asp	−2	−2	1	6	
cys	0	−3	−3	−3	9

often have negative scores. The largest scores are always in the diagonal of the
matrix, because there is no greater similarity than equality of two amino acids. When
multiple amino acids are aligned to a known sequence, the overall alignment score
of the sequence is simply the add-up sum of separate amino acid alignment scores.
There are many similarity score matrices today, but BLAST from the US National
Center of Biotechnology Information (National Institute of Health, Bethesda USA)
[11] mainly uses the BLOSUM 62 matrix, while MOTIF Search from the Biobase
database (Wolfenbüttel Germany) [12] uses a slightly different matrix, the Fasta
format. Both sites offer free analyses of novel sequences against known protein /
DNA sequences from their databases. BLAST has its own database, while MOTIF
Search applies 6 databases, e.g., PROSITE from Luzern Switzerland, BLOCKS
from Seattle USA, and PRINT from Manchester UK. The analysis-providers not
only account for similarity scores, but also assess the risk of deletions (as observed
in multiple data files so far), and account what will happen to the similarity scores,
if the best possible match is only observed with one or more amino acid deletions.
The effect of deletions, inserted by the providers' software, on the magnitude of the
similarity scores is, of course, that the latter will get smaller the more deletions are
inserted.

   Table 17.1 gives a part of the BLOSUM 62 matrix of similarity alignment scores
between individual amino acids (ala = alanine, arg = arginine, asn = asparagine,
asp = aspartic acid, cyst = cysteine). The similarity alignment scores are the
most important measures for testing new protein sequences against known protein
sequences from the database applied. For each alignment situation, e.g., $2 \times 2, 2 \times 3$,
2×4, 2×5, etc. amino acids, similarity alignment scores can be calculated. The
software makes a selection of alignment situations that produce large alignment
scores. We wish to test whether the alignment situations are statistically significant,
meaning that they occur more frequently than could happen by chance. This is
essential, because it means that the two sequences in the alignment situation are
more related to one another than could happen by chance, and, thus, that a biological
relationship (i.e., homology) is probable.

   To test whether an alignment situation is statistically significant, just assessing
that its score is larger than some kind of mean score based on known data is not
good enough. Instead, assessing that its score is better than an already previously
calculated large score is more adequate. Gumbel' s asymptotic distribution [14] for
extreme values is used for testing purposes, and uses the underneath equation:

$$f(x) = e^{-e^{-x}}$$

The expected number of statistically significant alignment scores for a given sequence is given by an E -term where

$$E \text{ is proportional to } mn.e^{-S}$$

and m and n are, respectively, the lengths of the unknown and known sequences and S is the add-up sum of multiple 2×1 amino acid alignments in an alignment situation given. Mathematically more convenient is the underneath equation.

$$E = K \, mn.e^{-\lambda S}$$

where K is a constant for adjusting the scoring matrix, like the BLOSUM 62 matrix used, and λ is the constant for the biochemical substance used (for example for protein 0.13). The E-term value is, actually, equal to the probability that an S value is observed by chance, and this probability is even adjusted for multiple testing, because it contains the multiple testing estimators m and n. And so, it can be considered a correct probability value (p-value), that the observed add-up similarity alignment score is not a chance finding, but, rather, a finding with a real meaning: a value < 0.05 means, that the null hypothesis of no difference from chance finding can be rejected. In BLAST high add-up similarity alignment scores are reported in a standardized way and called bit scores which are obtained by log-scaling the alignment scores according to:

$$Sbit = (\lambda.S - \log K)/\log 2$$

where Sbit is the bit score, the standardized score (the score that is independent of any unit), and log = natural logarithm. An example is given from the homepages of the University of Brussels [15]. If a bit score were 30, then you would have to assess on average $2^{30} = 1$ billion amino acid pairs. Think about your search space as the space for all candidate score solutions (here 0–1 billion). If our query protein has 235 amino acids and the database has 12,496,420 amino acids, then the adjusted search space of our data would normally be 0.13 times 235 times $12,496,420 = 0.38$ billion amino acids. Our observed search space would be much larger than expected, and, so, this difference is likely not to have arisen by chance. The magnitude of the probability that it would have arisen by chance can be calculated (Prob = probability):

$$\text{Prob of difference by chance} = K \, e^{-\lambda.S}$$
$$= K \, e^{(-\log 2.Sbit + \log K)}$$
$$= 2^{-Sbit}$$

This result must be multiplied with "mn" to adjust for multiple testing. To underscore the appropriateness of the above approach the following reasoning can be given. The probability of occurrences of "high similarity alignment scores" can be thought of as that of any random event to take place, and can thus be assumed to follow a Poisson distribution. Using this distribution, the underneath equation is adequate (Prob = probability).

$$\text{Prob of n occurrences of S} = (e^{-\mu}\mu^{n})/n!$$

(! = sign of faculty, e.g., 5! means 5.4.3.2.1, and μ equals the E-value of S). The chance of finding zero large similarity alignment scores with score higher than S is $e^{-\mu} = e^{-E}$. So, the chance of finding at least one such high alignment score situation is

$$1 - e^{-E}$$

This is the p-value of accepting the null hypothesis.

The BLAST program reports several terms:

1. Max score = best bit score between query sequence and database sequence.
2. Total score = best bit score if some amino acid pairs in the data have been used more often than just once.
3. Query coverage = percentage of amino acids used in the analysis.
4. E-value = expected number of large similarity alignment scores.

If the E-value is very small for the score observed, then a chance finding can be rejected. The sequences are then really related. An E-value = p-value adjusted for multiple testing = probability < 0.05. It indicates that the match between a novel and already known sequence is closer than could happen by chance, and that the novel and known sequence are thus homologous (phylogenetically from the same ancestor, whatever that means).

4.1   Example 1

We isolated the following amino acid sequence: serine, isoleucine, lysine, leucine, tryptophan, proline, proline. The one letter abbreviation code for this sequence is SIKLWPP. The BLAST Search site is explored, while giving the following commands.

```
Open BLAST Search site at appropriate address
     (Reference 1).
Choose Protein BLAST
Click Enter Sequences and enter the amino acid sequence
     SIKLWPP
Click BLAST
```

The output tables use the term blast hit which means here a database sequence selected by the provider's software to be largely similar to the unknown sequence, and the term query, which means here an unknown sequence that the investigator has entered for sequence testing against known sequences from the database. The output tables report

1. No putative conserved domains have been detected.
2. In the Distribution of 100 Blast Hits on the Query sequence all of the Blast Hits have a very low alignment score (< 40).
3. In spite of the low scores their precise alignment values are given next, e.g., the best one has a max score of 21.8, total score of 21.8, query coverage of 100 %, and adjusted p-value of 1,956 (not significant).

As a contrast search the MOTIF Search site is explored. We command.

```
Open MOTIF Search site at appropriate address
     (Reference 2).
Choose: Searching Protein Sequence Motifs
Click: Enter your query sequence and enter the amino
     acid sequence SIKLWPP
Select motif libraries: click various databases given
Then click Search.
```

The output table reports: 1 motif found in PROSITE database (found motif PKC_PHOSPHO_SITE; description: protein kinase C phosphorylation site). Obviously, it is worthwhile to search other databases if one does not provide any hits.

4.2   Example 2

We wish to examine a 12 amino acid sequence that we isolated at our laboratory, use again BLAST. We command.

```
Open BLAST Search site at appropriate address
     (Reference 1).
Choose Protein BLAST
Click Enter Sequences and enter the amino acid sequence
     ILVFMCWLVFQC
Click BLAST
```

The output tables report

1. No putative conserved domains have been detected.
2. In the Distribution of 100 Blast Hits on the Query sequence all of the Blast Hits have a very low alignment score (< 40).
3. In spite of the low scores their precise alignment values are given next. Three of them have a significant alignment score at $p < 0.05$ with max scores of 31.2, total scores 31.2, query cover of around 60 %, and E-values (adjusted p-values) of 4.1, 4.1, and 4.5. Parts of the novel sequence have been aligned to known sequences of proteins from a streptococcus and a nocardia bacteria and from

caenorhabditis, a small soil-dwelling nematode. These findings may not seem clinically very relevant, and may be due to type I errors, with low levels of statistical significance, or material contamination.

4.3 Example 3

A somewhat larger amino acid sequence (25 letters) is examined using BLAST. We command.

```
Open BLAST Search site at appropriate address
     (Reference 1).
Choose Protein BLAST
Click Enter Sequences and enter the amino acid sequence
     SIKLWPPSQTTRLLLVERMANNLST
Click BLAST
```

The output tables report the following.

1. Putative domains have been detected. Specific hits regard the WPP superfamily. The WPP domain is a 90 amino acid protein that serves as a transporter protein for other protein in the plant cell from the cell plasma to the nucleus.
2. In the Distribution of 100 Blast Hits on the Query sequence all of the Blast Hits have a very high alignment score (80–200 for the first 5 hits, over 50 for the remainder, all of them statistically very significant).
3. Precise alignment values are given next. The first 5 hits have the highest scores: with max scores of 83.8, total scores of 83.8, Cover queries of 100 %, and p-values of $4\,e^{-17}$, which is much smaller than 0.05 (5 %). All of them relate to the WPP superfamily sequence.

   The next 95 hits produced Max scores and Total scores from 68.9 to 62.1, query coverages from 100 % to 96 % and adjusted p-values from $5\,e^{-12}$ to $1\,e^{-9}$, which is again much smaller than 0.05 (5 %).
4. We can subsequently browse through the 95 hits to see if anything of interest for our purposes can be found. All of the alignments as found regarded plant proteins like those of grasses, maize, nightshade and other plants, no alignments with human or veterinary proteins were established.

4.4 Example 4

A 27 amino acid sequence from a laboratory culture of pseudomonas is examined using BLAST. We command.

```
Open BLAST Search site at appropriate address
     (Reference 1).
```

```
Choose Protein BLAST
Click Enter Sequences and enter the amino acid sequence
    MTDLNIPHTHAHLVDAFQALGIRAQAL
Click BLAST
```

The output tables report

1. No putative domains have been detected.
2. The 100 blast hit table shows, however, a very high alignment score for gentamicin acetyl transferase enzyme, recently recognized as being responsible for resistance of pseudomonas to gentamicin. The ailments values were a max score and total score of 85.5, a query coverage of 100 %, and an adjusted p-value of $1 e^{-17}$, and so statistically very significant.
3. In the Distribution of the 99 remaining Blast Hits only 5 other significant alignment were detected with max score and total scores from 38.5 to 32.9, query coverages 55–92 %, and adjusted p-values between 0.08 and 4.5 (all of them 5 %). The significant alignments regarded bacterial proteins including the gram negative bacterias, rhizobium, xanthomonas, and morganella, and a mite protein. This may not clinically be very relevant, but our novel sequence was derived from a pseudomonas culture, and we know now that this particular culture contains pathogens very resistant to gentamicin.

5 Discussion

Mendel started a novel science that now 140 years later is the largest growing field in biomedicine. This novel science, although in its first steps, already has a major impact on the life of all of us. E.g., obtaining enough drugs, like insulin and many others, to treat illnesses worldwide was a problem that has been solved by recombinant DNA technology which enabled through genetic engineering of bacteria or yeasts the large scale production of various pharmaceutical compounds. The science of genes, often called genomics, is vast, and this chapter only briefly mentions a few statistical techniques developed for processing data of genetic research.

Although high throughput methods are still relatively expensive, and are not used routinely in clinical trials, these methods undoubtedly will be used more often in the future. Their promise of identifying subgroups of patients with varying drug response is of major importance and is a major topic of pharmaco-genomics. In addition, differential expression profiles, and proteomics are of major importance of identifying new pathways for targeting new drugs. More sophisticated statistical methods are required, and will be developed.

With the help of gene expression analyses thousands of genetic variants and differentially expressed genes have been associated with clinical effects / functions, and, in addition, the proteins associated with these genes have been investigated to disentangle their roles in the biochemical and physiological pathways of the disease and treatments being studied. Sequence similarity searching is a method that can be applied by almost anybody for finding similarities between his / her query sequences and the sequences known to be associated with different clinical effects.

With sequence similarity searching the use of p-values to distinguish between high and low similarity is relevant. Unlike the BLAST [11] interactive website, the MOTIF [12] interactive website does not give them, which hampers inferences from the alignments to be made.

6   Conclusions

1. With the help of gene expression analyses thousands of genetic variants and differentially expressed genes have been associated with clinical effects / functions, and, in addition, the proteins associated with these genes have been investigated to disentangle their roles in the biochemical and physiological pathways of the disease and treatments being studied.
2. Sequence similarity searching is a method that can be applied by almost anybody for finding similarities between his / her query sequences and the sequences known to be associated with different clinical effects.

References

1. Cornelisse CJ, Cornelis RS, Devilee P (1996) Genes responsible for familial breast cancer. Pathol Res Pract 192:684–693
2. Wijnen JT, Vasen HF, Khan PM, Zwinderman AH, van der Klift H, Mulder A, Tops C, Moller P, Fodde R (1998) Clinical findings with implications for genetic testing in families with clustering of colorectal cancer. N Engl J Med 339:511–518
3. Jordan B (ed) (2001) DNA microarrays: gene expression applications. Springer, Berlin
4. Claverie JM (2001) Computational methods for the identification of differential and coordinated gene expression. Hum Mol Genet 8:1821–1832
5. McLachlan G (2001) Mixture.model clustering of microarray expression data. In: Aus biometrics and New Zealand Stat Association Joint conference, Christchurch
6. Alizadeh AA et al (2000) Distinct types of diffuse large B-cell lymphoma identified by gene expression profiling. Nature 403:503–511
7. Eisen M et al (1998) Cluster analysis and display of genome-wide expression patterns. Proc Natl Acad Sci USA 95:14863–14867
8. Tavazoie S et al (1999) Systematic determination of genetic network architecture. Nat Genet 22:281–285
9. Tamayo P et al (1999) Interpreting patterns of gene-expression with self-organizing maps. Proc Natl Acad Sci USA 96:2907–2912
10. Tibshirani R et al (2005) Clustering methods for the analysis of DNA microarray data. Technical Reports Stanford University, Department of Statistics, Stanford
11. BLAST (2013) http://blast.ncbi.nlm.nih.gov/Blast.cgi. 13 Mar 2013
12. MOTIF Search (2013) http://www.genome.jp/tools/motif. 13 Mar 2013

13. Dayhoff MO, Schwarz RM, Orcutt BC (1978) A model of evolutionary change in proteins. In: Dayhoff MO (ed) Atlas of protein sequence and structure, vol 5, Suppl 3. National Biomedical Research Foundation, Washington, DC, pp 353–358
14. Anonymous (2013) 1.3.6.6.16. extreme value type 1 distribution. In: NIST/SEMATECH e-handbook of statistical methods. http://www.itl.nist.gov/div898/handbook/index2.htm. 13 Mar 2013
15. Anonymous (2013) Score, bit-score, P-value, E-value. www.homepages.ulb.ac.be/~dgonze/TEACHING/stat_scored.pdf. 14 Mar 2013

Chapter 18
Continuous Sequential Techniques

1 Summary

1.1 Background

The basic principle of sequential statistical testing is that after every additional test person has been evaluated, a formal statistical rule is applied to the whole data to determine whether the study should stop. This methodology has obtained its proper place in econo- and sociometric research as a cost-efficient and scientifically sound methodology for efficacy/safety testing of industrial products, but is underused in clinical research.

1.2 Objective

To discuss and review possibilities for including in clinical research sequential testing for proportions, paired and unpaired continuous data.

1.3 Methods

Using real and hypothesized examples, we will assess both parallel and triangular sequential analysis sets.

1.4 Results

The sequential probability ratio test of Wald produces parallel linear stopping boundaries, and could go on infinitely unless a maximal sample size is defined in the

T.J. Cleophas and A.H. Zwinderman, *Machine Learning in Medicine:*
Part Two, DOI 10.1007/978-94-007-6886-4_18,
© Springer Science+Business Media Dordrecht 2013

protocol. It is adequate for comparing proportions like numbers of patients cured or not. In the programs for sequential comparisons of paired and unpaired differences as developed by Armitage and Whitehead the rejection and acceptance boundaries are linear too, but tend to converge. This is convenient, because a maximal sample size need not be defined.

1.5   Conclusions

1. Sequential trials have precise stopping rules preventing ethical, financial and scientific problems.
2. Sequential trials are not necessarily biased, although the risk of gambler's bias is possible.
3. The statistical software of sequential trials is pretty expensive, but basic tests can be performed using a pocket calculator.
4. Sequential trials involve multiple testing, and, therefore, require Bonferroni adjustments.
5. Sequential trials are at a risk of early stop, and, therefore power loss due to small samples.

2   Introduction

Historically, the statistical theory for stopping rules in clinical trials has been largely concerned with sequential designs for continuous monitoring of treatment differences. The basic principle is that after every additional patient on each treatment has been evaluated, some formal statistical rule is applied to the whole data so far, to determine whether the trial should stop. The theory of sequential techniques is already quite old. It was developed in the early forties [1]. At that time there was massive increase in industrial production. There was a need to ensure that these products, including munitions for the war industry, were reliable, requiring methods of minimal testing, with yet reliable results. To meet these requirements Wald [2] of the Ministry of Supply, Washington DC, USA, developed his sequential probability ratio test. Independent of him Alan Turing [3] at Bletchley Park, Bletchley, UK, developed similar techniques to test hypotheses about deciphering encrypted messages coded by the German Enigma machines in the Second World War. At this time the methodology of continuous sequential statistical techniques has obtained its proper place in econo- and sociometric research as a cost-efficient and scientifically sound methodology for efficacy/safety testing of industrial products, but is underused in clinical research [4–6].

The central idea is to calculate after each additional patient (a function of) the treatment difference, and the total amount of information, sampled thus far. These two statistics are plotted graphically against each other each time a new patient is evaluated. The stopping rule of the trial entails evaluating whether a boundary is crossed. The upper boundary represents the threshold for rejecting the null hypothesis, the lower boundary that for accepting the null hypothesis. The two boundaries may be parallel but this is not necessarily so. We will discuss and review possibilities for including in clinical research sequential testing for proportions, paired and unpaired continuous data.

3 Parallel and Triangular Sequential Sets

Figure 18.1 gives an example of the sequential probability ratio test invented by Wald [2] and applied as a criterion for industrial manufacturing, psychological testing and classifying academic examinees [7]. In clinical research it was used as a stopping criterion in phase III clinical trials [4,5], and continuous safety monitoring [6]. In therapeutic research, a patient may be positive or negative (e.g., cured or not). The y-value gives the cumulative number of positive patients, n is the total number of tests with, at each test, another patient added to the sample. The upper oblique line represents the border for rejecting the null hypothesis of no significant difference in the data, the lower oblique line represents the border for accepting the null hypothesis. The two borders are, obviously, parallel. It can be observed, that, after 12 patients have been added, a significant difference is observed, and, that the null hypothesis of no difference in the data can be rejected. The study is stopped.

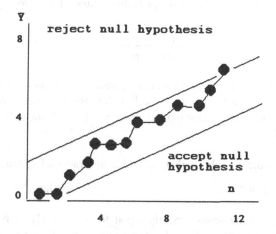

Fig. 18.1 Example of a sequential probability ratio test: y = the cumulative number of positive tests, n is the total number of tests with after each test adding a subsequent patient. The upper oblique line represents the border for rejecting the null hypothesis of no significant difference in the data, the lower oblique line represents the border for accepting the null hypothesis. It can be observed that after 12 patients have been added to the sample a significant difference is observed and that the null hypothesis of no difference in the data can be rejected. The study is stopped

We should emphasize, that, like any other statistical method involving multiple testing, sequential testing requires that p-values be adjusted either using Bonferroni adjustment or another method.

Wald's sequential probability ratio test is adequate for testing ratios of proportions or means. For sequentially testing differences between paired comparisons and unpaired comparisons Armitage [8] and Whitehead [9] developed alternative methods. All of them involve linear borders for accepting null and alternative hypotheses, and borders tend to have positive direction coefficients. This is, because, the larger the number of tests, the larger the cumulative number of positive patients. With the Armitage and Whitehead models, unlike the Wald's model, the boundaries of uncertainty, however, tend to converge. This can be explained by the Martingale variance of the central limit theory: the uncertainty of stochastic processes with continuous data tend to converge with large samples [10].

4   Data Example of a Parallel Sequential Set

A data example is given using the Wald sequential probability ratio test. The mean half life of a drug should be less than 20 h (null hypothesis H_0) and better than 10 h (H_1). We test the drugs every 2 h to assess whether they are still active. The drug lives are exponentially distributed.

The probability that the drug survives the first test under H_0 is $e^{-2/20} = 0.905$, under H_1 it is $e^{-2/10} = 0.819$.

Using the binomial distribution, the probability ratio of the two hypotheses is

$$\frac{p_1{}^y (1 - p_1)^{n+y}}{p_0{}^y (1 - p_0)^{n+y}}$$

where n is number of tests, y is number of successful tests, and p_0 and p_1 are the probabilities that the drug survives. For defining the two hypothesis errors alpha = type I error and beta = type II error must be defined. In this example alpha = beta = 0.128.

Beta = the probability that the probability ratio is large than A,

1- alpha = ""   "   "   smaller " B,

Where A and B are approximately equal to respectively (1-beta)/alpha and beta/(1-alpha), which are 0.147 and 6.813 respectively, A and B can be used to decide either to accept H_0 or H_1 or to continue testing, that is if

B < probability ratio < A

i.e.,

$$0.147 < \text{probability ratio} < 6.813.$$

An example is given: with $n = 10$ and $y = 6$ the probability ratio is computed

$$\frac{0.819^6(1 - 0.819)^{(10-6)}}{0.915^6(1 - 0.915)^{(10-6)}} = 7.23$$

The inequality "B < probability ratio < A" can be more conveniently solved by logarithmic transformation according to

$$\ln B < (an + by) < \ln A$$

where $a = \ln(1 - p_1)/(1 - p_0)$ and $b = [\ln(p_1/p_0) - \ln(1 - p_1)/(1 - p_0)$.

In the given example

$$a = 0.6446$$

$$b = -0.7444.$$

and

$$\text{Ln } B = \ln 0.147 = -1.919$$

$$\text{Ln } A = \ln 6.183 = 1.919$$

With $n = 10$ and $y = 6$ we find that

$$(an + by) = (0.64446 \times 10 - 0.7444 \times 6) = 1.979.$$

This result is both larger than -1.919 and 1.919. We can stop testing and accept H_1. The drug performs significantly worse than 10 h.

A more convenient way to describe the above procedure is using the underneath equations that can be derived from the above data example.

$$\text{Ln } B < (an + by) < \ln A$$

$$(\ln B/b - an/b) < y < (\ln A/b - an/b)$$

$$(2.578 + 0.8666n) > y > (-2.578 + 0.8666n)$$

From the above follows:

$$y = 2.578 + 0.8666\,n$$

$$y = -2.528 + 0.8666\,n$$

The two equations can be used for describing the linear stopping boundaries while the region in-between is the region of continuation of the trial (compare Fig. 18.1).

The equations have the same direction coefficients, and, so, the lines are parallel. As long as they are not crossed, the sequential trial could go on infinitely, unless some form of maximal n is included in the protocol.

5 Triangular Sequential Sets

In Fig. 18.2 a triangular sequential set is given. After each additional patient (a function of) the treatment difference, in this example called Z, and the total amount of information, in this example called V, are calculated. These two statistics are plotted graphically against one another each time a new patient is evaluated. The stopping rule of the trial entails evaluating whether a boundary is crossed. A typical example of a sequential trial with a so-called triangular sequential set is illustrated.

The undulating line illustrates a possible realisation of a clinical trial: after each time a new patient could be evaluated, Z and V are calculated and the line is extended a little further. The line-sections AC and BC are the stopping boundaries, and the triangular region ABC is the continuation region. If the sample path crosses AC, the null hypothesis will be rejected at the 5 % significance level, and if BC is crossed then H_0 will be accepted.

The triangular test is one of many possible sequential trial designs; but the triangular test has some very attractive characteristics. If the treatment difference is large, it will lead to a steeply increasing sample path, and, consequently, to a small

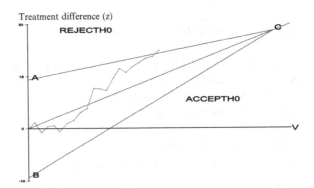

Fig. 18.2 Typical example of a sequential trial with a so-called triangular test. The undulating line illustrates a possible realisation of a clinical trial: after each time a new patient could be evaluated, Z (*variable vertical axis*) and V (*variable horizontal axis*) are calculated and the line is extended a little further. The line-sections AC and BC are the stopping boundaries, and the triangular region ABC is the continuation region. If the sample path crosses AC, the null hypothesis is rejected at the 5 % significance level, and if BC is crossed then H_0 is accepted

trial, because the AC boundary is reached rapidly. If there is no difference between treatment, the sample path will move horizontally and will cross the BC boundary quickly which also leads to a small trial. If the treatment difference is negative, the BC boundary will be crossed even quicker.

The trick is to devise sensible boundaries. Whitehead [11] gives an elaborate discussion on how to do this (as well as how to calculate Z and V). She also discussed many different sequential plans for many different types of clinical trials and data-types. Whitehead and her associates have also developed a user-friendly, although expensive ($1,000 per trial) computer program to design and analyze sequential clinical trials [11]. However, basic sequential testing like the Wald sequential probability test, can be readily performed with the help of a pocket calculator as observed in the example of Sect. 4.

6 Discussion

Unlike traditional trials that have one or two rather arbitrary interim analyses, sequential trials have precise stopping rules that can prevent several design problems.

1. The ethical problem of too many patients giving a potentially inferior treatment,
2. The scientific problem of requiring the repetition of research after a negative trial,
3. The financial problem of involving the costs with too small or too large studies.

Sequential trials are widely applied in econo- and sociometric research, but are underused by clinical investigators [4–6], maybe because of their traditional belief in fixed sample sized randomized trials. This is unfortunate considering the above risks with traditional trial designs.

We should add that terminating a trial early is not necessarily a biased activity. Several statisticians have demonstrated that results, although they are at a loss of accuracy due to loss of statistical power, are, indeed, not biased [12].

However, the decisions either to proceed or stop sequential testing are, essentially, one direction decisions, and may, therefore, be at risk of a special type of bias, called gambler's fallacy: if you increase your bet after you have won, and never decrease it you will end up broke [12]. However, if in a dataset the priori chance of a second direction is clinically very small, this potential bias may not have to be taken into account.

The sequential probability ratio test of Wald [2] produces parallel linear stopping boundaries, and could go on infinitely unless a maximal sample size is defined in the protocol. It is adequate for comparing proportions like numbers of patients cured or not. In the programs for sequential comparisons of paired and unpaired differences as developed by Armitage [8] and Whitehead [9] the rejection and acceptance boundaries are linear too, but tend to converge. This is not inconvenient, because the definition of a maximal sample size is no further needed.

Sequential trials, like any other trials involving multiple testing, require p-value adjustment for multiple testing, e.g., Bonferroni adjustment. Sequential trials are also at risk of power loss due to a reduced sample size when the study is stopped early. This has to be taken into account.

7 Conclusions

1. Sequential trials have precise stopping rules preventing ethical, financial and scientific problems.
2. Sequential trials are not necessarily biased, although the risk of gambler's bias is possible.
3. The statistical software of sequential trials is pretty expensive, but basic tests can be performed using a pocket calculator.
4. Sequential trials involve multiple testing, and, therefore, require Bonferroni adjustments.
5. Sequential trials are at a risk of early stop, and, therefore power loss due to small samples.

References

1. Demets DL, Ware JH (1980) Group sequential methods in clinical trials with a one sided hypothesis. Biometrika 67:651–660
2. Wald A (1945) Sequential tests and statistical hypotheses. Ann Math Stat 16:117–186
3. Hodges A (1983) Alan Turing: the Enigma. Burnett Books, London
4. Bogowicz P, Flores-Mir C, Major PW, Heo G (2008) Sequential analysis applied to clinical trials in dentistry: a systematic review. Evid Based Dent 5:55–60
5. Todd S, Whitehead A, Stallard N, Whitehead J (2010) Interim analyses and sequential designs in phase III studies. Br J Clin Pharmacol 51:394–399
6. Ball G, Silverman MH (2011) Continuous safety monitoring for randomized controlled trials with blinded treatment information. Comtemp Clin Trials 32(S1):S8–10
7. Anonymous (2013) Sequential probability ratio test. http://en.wikipedia.org. 12 Feb 2013
8. Armitage P (1975) Sequential medical trials. Blackwell, Oxford
9. Whitehead J (1983) The design and analysis of sequential clinical trials. Ellis Horwood, Chichester
10. Lai TL (2004) Martingales in sequential analysis and time series, 1945–1985. Research founded by the National Science foundation, grant DMS 0805879. lait©stat.stanford.edu. 12 Feb 2013
11. Whitehead J (1998) Planning and evaluating sequential trials (PEST, version 3). University of Reading, Reading. www.reading.ac.uk/mps/pest/pest.html. 12 Feb 2013
12. Anonymous (2013) Why is bias affected when a clinical trial is terminated at an early stage? http://stats.stackexchange.com. 10 Feb 2013

Chapter 19
Discrete Wavelet Analysis

1 Summary

1.1 Background

Wavelets are oscillations, supposedly resulting from multiple smaller wavelets, and they are, traditionally, analyzed with polynomial, sine and cosine, and other functions. Ingrid Daubechies (1988) demonstrated that the repeated use of sharply spiked functions with multiple scales as basis functions for wavelet analysis provided better data-fit, and called it discrete wavelet analysis.

1.2 Objective

In medicine discrete wavelet analysis might be useful for the study of changes across time, but is little used so far.

1.3 Methods

Using three real data examples, (1) an electrocardiogram, (2) a daily blood glucose curve, (3) and a health institution's daily energy consumption, we assessed whether discrete wavelet analysis was adequately precise for describing wavelike patterns in therapeutic research and health care.

T.J. Cleophas and A.H. Zwinderman, *Machine Learning in Medicine:*
Part Two, DOI 10.1007/978-94-007-6886-4_19,
© Springer Science+Business Media Dordrecht 2013

1.4   Results

An excellent fit of discrete wavelet analyses was provided for all of the three examples with means and standard deviations virtually 100 % identical to those of the original data.

1.5   Conclusions

1. None of the traditional models for describing non linear data fits wavelike data so well as discrete wavelet analysis does. This is, because (1) sharply spiked prototype wavelets appropriate to the data are used, (2) tiny fluctuations are measured at the appropriate scales, and (3) families of functions have an orthogonal basis.
2. Unlike traditional non linear models, it is sparsely coded, and, therefore, requires relatively little computer memory.
3. Unlike many traditional non linear models, it is able to differentiate noise from real effects.
4. Tiny fluctuations can be recognized as real effects.
5. Unlike traditional regression models, it is appropriate for constantly changing data, like electrocardiograms, and health institutions' consumption patterns.
6. A disadvantage is that the excellent fit is accompanied by a lacking mathematical equation to describe the assumed relationship with time. Such an equation is convenient for making predictions. However predictions based on ill-fitting mathematical equations do not mean too much.

2   Introduction

Wavelets are brief oscillations like the recordings of an electro-cardiograph (ECG), -encephalograph, -myograph, phonocardiograph, respiratory flow meter, continuous ambulatory blood pressure recorder etc. [1–3]. They, essentially, start at mean (zero), increase and/or decrease, and then return to mean. Discrete wavelet analysis may provide a more precise fit of time series than traditional regression modeling like polynomial and Fourier analysis [1–3].

Generally, a wavelet as observed is considered to be the resultant of multiple smaller wavelets, like an ECG recording measured at the body surface but resulting from electrical changes in the two small and two large heart chambers [4]. In order to better understand the pattern of an observed wavelet, it would, therefore, make a lot of sense to try and decompose it into fragments. Already in the first part of the nineteenth century the French mathematician Joseph Fourier [5] used this concept, when he developed a method for representing a wavelet with an analysis function constructed as the linear sum of multiple sine and cosine functions. Since the early

30th several investigators have been applying more sharply spiked functions instead of sines and cosines as basis functions for wavelet analysis, and have added their names to the wavelet functions they invented like Haar [6], Daubechies [7], Morlet [8], Coifman [9] etc. However, wavelet analysis made a tremendous progress after Ingrid Daubechies (1988) [7] proposed the repeated use of a single basis function with multiple scales. This method is now available not only in commercial statistical software programs like S-plus [10], and Matlab [11], but also as a free software and wavelet calculator on the Internet, e.g., the interactive wavelet program from Ion.researchsystems [12]. Discrete wavelet analysis is currently applied in many more fields than oscillography, e.g., imaging and image compression, de-noising noisy data, quantum physics, electrical engineering, etc. Some researchers even feel that by using it, one is adopting a whole new mindset in data processing [2].

In medicine discrete wavelet analysis might be useful for the evaluation of treatment effects across time not only in individuals, but also in therapeutic research and health care at a broader level. Yet it is little used in these fields so far [13]. This chapter, using real data examples, was written to assess whether discrete wavelet analysis is adequately precise for the purpose. We will give step-by-step analyses for the convenience of the readers.

3 Some Theory

Fourier's (1807) proposed construct for wavelet modeling was the following function [5].

$$f(t) = p + q_1 \cos(t) + .. + q_n \cos n(t) + r_1 \sin(t) + .. + r_n \sin n(t)$$

with p, $q_1 \ldots q_n$, and $r_1 \ldots r_n$ = constants for the best fit of the given data, and t = time. This analysis fits wavelets with smooth sinusoidal patterns very well, but with discontinuities and multiple sharp spikes this is less so, and mathematicians have wanted more appropriate functions for their analysis [1–3]. Already in the early 30th several groups working independently researched the possibility to analyze wavelets using scale-varying basis functions. A single low frequency basis function was used as the prototype wavelet, called the mother-wavelet, and many similar functions although at smaller scales, called the father- and baby-wavelets were included to produce the best fit model of the wavelet as observed. Figure 19.1 gives an example of a prototype wavelet with a single baby wavelet of a different amplitude and scale, but, otherwise, of the same contour. Just like with Fourier analysis [5], multiple baby wavelet functions with an endless number of scales can be used to describe the original wavelet as measured in the best possible way. Mathematically the analysis function can be described as follows.

$$f(t) = f_{mother}(t) + f_{father}(t) + f_{baby1}(t) + f_{baby2}(t) + f_{baby3}(t) + \ldots \ldots$$

Fig. 19.1 A prototype
wavelet named after its
inventor Mrs. Ingrid
Daubechies (the baby wavelet
has the same contour as the
mother wavelet, but a
different amplitude and
frequency (= 1/scale))

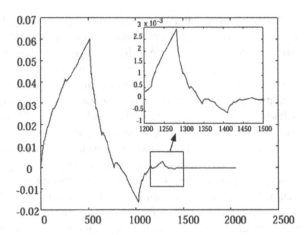

Often the mother function is used to describe the largest low frequency fragment
of the wavelet measured, while the father function is the largest spike of it. The baby
functions are for refinement. The baby functions are, thus, identical to mother and
father functions, but they are shifted along the time axis by some amount and the
scale is changed for an accurate fit.

There are some very special advantages of the scale-varying basis functions
compared to Fourier analysis [11]. First, not only does it better fit wavelets with
sharp spikes, but, second, also it is less sensitive to noise, because it is able to
measure tiny fluctuations in the original wavelet at the appropriate scales. A third
advantage is that the family of functions used for the contract has an orthogonal
basis. It means that the correlation levels between the paired functions does not
have to be taken into account, leading to non-redundancy, i.e., minimal residuals,
otherwise called, minimal uncertainty in the models. This is again a great advantage
with complex wavelets and large data, because it can be sparsely coded, and,
thus, requires relatively little computer memory. This is not so with traditional
methods potentially requiring computations of weeks or months even on modern
computers. Discrete wavelet analysis has no problems with that, and can even
be used for data compression with virtually no loss of information. Because it is
able to measure tiny fluctuations in the original wavelet, it is also better able than
traditional methods to recognize noise due to, e.g., a transiently defect recording
device. Universal or global denoising, otherwise called shrinkage, makes use of the
2 std (standard) deviation criterion: small wavelets that cross the 2 std deviation
boundary of their areas under the curve are removed. This can be done without
appreciable degradation or loss of information.

4   Clinical Relevance of Discrete Wavelet Analysis

In clinical medicine one might think of applying discrete wavelet analysis not only of physical changes in the body, like, e.g., blood pressure changes, but also of changes in chemical blood values like glucose levels in individuals or groups, fluctuations in energy or health means consumptions in institutions or countries and many more types of data.

As a first example, the electrocardiogram of a single heart beat will be assessed for features in the recording that are important and those that can be ignored. For the purpose a two dimensional discrete wavelet power spectrum using a Morlet wavelet will be applied [8].

As a second example, we will assess the 1,100 glucose measurements in 3 years of one diabetic patient with a stable body weight and treatment. A weak positive correlation of glucose levels with temperature, and amount of sunlight has sometimes been reported [14], and a weak negative correlation with age [15]. Discrete wavelet analysis using the Coifman 1 wavelet [9] will be used as a very precise method to evaluate the eventual presence of such phenomena.

As a third example, energy consumption in a large institution like a hospital is studied for trends and patterns. Energy consumption in health institutions has rocketed due to automation and the use of electronic patient files [16]. Also monitoring health care consumption, including drug consumption/clinical research costs, are important issues, especially today [16]. The Daubechies 1 and 3 wavelets [7] will be used for analysis. The statistics of the precision after reconstruction and global denoising of the original energy consumption pattern are given.

5   First Example, a Single Heart Beat's Electrocardiogram

First, the electrocardiogram of a single heart beat is assessed for features in the recording that are important and those that can be ignored. For the purpose a two dimensional discrete wavelet power spectrum using a Morlet wavelet will be applied [8]. A free software and wavelet calculator on the Internet offering an interactive wavelet program from Ion.researchsystems was used for analysis [12].

The wave's amount of energy at a certain point of time is equivalent to its amplitude-squared. A discrete wavelet transform is obtained by cutting the observed wave into several fragments with different frequencies (e.g., the entire wave has a frequency very different from those of the small sub-waves like the P, QRS and T waves, that are of much shorter duration), and the Morlet prototype wavelet and its baby wavelets are used for the transformation. A wavelet power spectrum is produced by the software as a two-dimensional picture, with time on the x-axis and the wavelet scales (1/frequencies) is on the y-axis.

> Start Google-Windows . . . .Enter: ION Script Wavelet . . . .Interactive Wavelets . . . .Visualization Options . . . .Enter your own data . . . .Data values box: enter your data . . . .Submit.

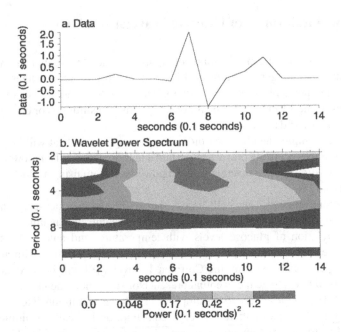

Fig. 19.2 The initial signal (*upper graph*) and the wavelet power spectrum calculated by the discrete wavelet program (*lower graph*). The density levels of the different areas indicate respectively that 5 %, 25 %, 50 %, and 75 % of the wavelet power is above each level. At, e.g., 0.7 s, it can be observed, that the discrete wavelet transform wave contains many high frequency and few low frequency wavelets: over 75 % of the power is in the 0.2–0.3 s periods (y-axis) and less than 5 % of it is in the 0.8 or more seconds periods (y-axis). In the other areas only lower power is generated. We can conclude that the interval 0.6–0.8 s (the QRS interval) contains the features of our signal most important

Figure 19.2 gives the graph of the wavelet power spectrum as produced. At, e.g., 0.7 s, it can be observed, that the discrete wavelet transform wave contains many high frequency and few low frequency wavelets: over 75 % of the power is in the 0.2–0.3 s periods (y-axis) and less than 5 % of it is in the 0.8 or more seconds periods (y-axis). In the other areas only lower power is generated. We can conclude that the interval 0.6–0.8 s (the QRS interval) contains the features of our signal most important and that the remainder can be largely ignored here.

6   Second Example, Trends and Patterns in a Patient's Blood Glucose Levels

The 1,100 subsequently measured glucose levels in 3 years of one diabetic patient with a stable body weight and treatment are in Fig. 19.3. The overview shows a wild signal. A weak correlation with temperature, sunlight, and age has been

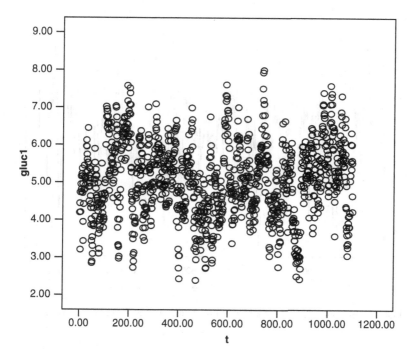

Fig. 19.3 1,100 glucose measurements from one patient

reported in previous publications [14, 15], but these features are not evident from the Fig. 19.3. Discrete wavelet analysis using S-plus statistical software [10] is applied as a very precise method for evaluating the presence of such phenomena in this patient with the help of a Coifman 1 (Coiflet 1) prototype wavelet and a three-level decomposition of baby wavelets, the reconstructed signal produces a wavelet with considerable noise (Fig. 19.4). We denoise the reconstructed wavelet using universal or global denoising. Figure 19.5 (upper graph) gives the denoised signal together with the original values. Figure 19.6 the upper graph plots the original values against the denoised values from the Coiflet discrete wavelet analysis. A perfect fit is observed. The result, thus, seems to be adequately precise for the purpose of making clinical inferences. In Fig. 19.5 (lower graph) a slightly positive overall trend of time is suggested. However, linear regression with time as predictor and glucose levels as outcome was computed not to be statistically significant. In contrast, a linear regression with the negative peak values from the discrete wavelet analysis as outcome was statistically significant ($r = 0.741$, r-square $= 0.549$, $F = 8.5$, $p = 0.022$). Also, in the months of January with little sunshine and low temperatures the glucose levels tended to peak, as indicated by the three small circles in Fig. 19.5 (lower graph).

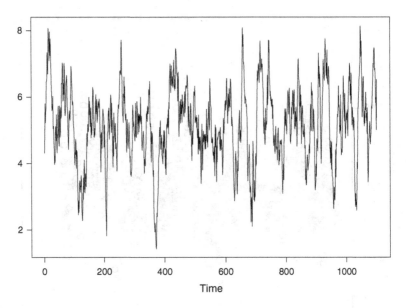

Fig. 19.4 Values from Fig. 19.3 according to discrete wavelet transform (S-plus software)

7   Third Example, Health Institutions' Ongoing Consumption Patterns

A health institutions' consumption pattern may show irregularities, and, in addition, may be sometimes noisy due to transient defects of the monitoring. Discrete wavelet analysis is better suitable than traditional methods for such features as explained in the above section "Some theory". A real data example, also used by Matlab [11] in its demo entitled Leleccum might serve as an example. The software program of MATLAB including the demo can be downloaded freely for the first month's of use. A stepwise analysis using the Menu program requires the following commands.

> From Matlab prompt type: wavemenu....click wavelet 1-D....File menu load signal....select demo: leleccum.mat....from wavelet choose db1 wavelet (Daubechies 1)....level: 3....click Analyze....click De-noise ....click Fixed form threshold....click De-noise again....View Denoised Signal....Close....Close De-noising window....Update synthesized signal: click no....Refine analysis: click db 3....click level 5....Analyze....click Compress

The "retained energy" summary tells that the compression process removed most of the noise, but preserved 99.97 % of the energy of the signal. Automated thresholding was very efficient zeroing out only 3.2 % of the wavelet frequencies. We continue the analysis clicking "residuals".

> Click Residuals....click respectively (1) original signal, (2) approximation, (3) synthesized signal.

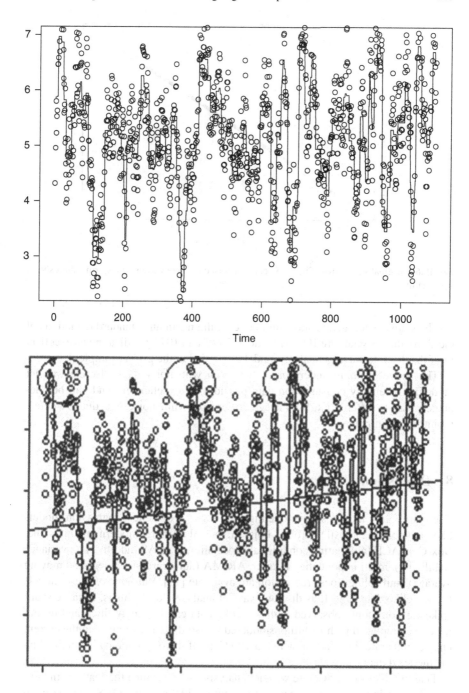

Fig. 19.5 *Upper graph*: the denoised signal from Fig. 19.4 after global thresholding with original values added (S-plus software) *Lower graph*: A slightly positive overall trend of time is suggested. A linear regression with time as predictor and the negative peak values of the discrete wavelet analysis as outcome was statistically significant ($r = 0.741$, r-square $= 0.549$, $F = 8.5$, $p = 0.022$). Also, in the months of January with little sunshine and low temperatures the glucose levels tended to peak as indicated by the three small *circles*

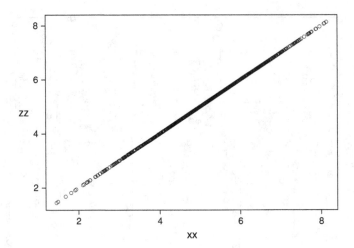

Fig. 19.6 Original values from Fig. 19.5 versus denoised discrete wavelet transform values shows a perfect fit

The statistics are given, showing that, e.g., the mean and standard deviations of the above three signals are 100 % identical (339.5 and 107.7), indicating an excellent fit of the discrete wavelet analysis model as applied in the given example.

The above example underscores, that, even with noisy data, discrete wavelet modeling of energy consumption data provides an excellent fit, and can be adequately used to differentiate brief irregularities from white noise, e.g., due to a defect monitoring equipment.

8 Discussion

There are many methods for modeling time series. Non linear regression models like polynomial, logistic, probit, Poisson, but also more modern methods like Box-Cox, ACE (alternating conditional expectations), AVAS (additive and variance stabilizing), multi-exponential, Fourier, ARIMA (autoregressive integrated moving average), and spline/Loess regression are available [17]. But none of these models fits wavelike data as well as discrete wavelet analysis does. This is, because sharp spikes as commonly observed with wavelike data can be nicely fitted and do not have to be replaced with ill-fitting smoothed sinusoids. The examples in the current chapter illustrate this, and show that a 100 % fit of the original data is obtained by the modeled data.

Due to the close fit discrete wavelet analysis can determine the features in your data that are important and those that are not so, like e.g., the QRS interval in an electrocardiogram.

A disadvantage is that the excellent fit is accompanied by the lack of a (simple) mathematical equation to describe the assumed nature of the relationship with time. Such an equation is convenient for making predictions about future data and future risk profiling in individuals. However, simple mathematical equations with ill-fit and wide standard errors mean little, due to the uncertainty they provide. In contrast, the perfectly fitting discrete wavelet technology enables to measure tiny fluctuations in the original wavelet with virtually 100 % certainty, and, in addition, enables to tell them apart from noise. This also means that all of the properties of the reconstructed wavelet can be interpreted as real effects (with a real meaning). Another advantage of discrete wavelet analysis is, that it is appropriate for data that change constantly, like e.g., electrocardiograms, and energy consumption data.

9 Conclusions

1. None of the traditional models for describing non linear data fits wavelike data as well as discrete wavelet analysis does. This is because sharply spiked prototype wavelets appropriate to the data are used, tiny fluctuations are measured at the appropriate scales, and families of functions have an orthogonal basis, meaning that they can be considered independent of one another.
2. Unlike traditional non linear models, it is sparsely coded, and, thus, requires little computer memory.
3. Unlike many traditional non linear models, it is able to differentiate noise from real effects.
4. Tiny fluctuations can be interpreted as real effects.
5. Unlike traditional regression models, it is appropriate for constantly changing data, like electrocardiograms, and energy consumption patterns.
6. A disadvantage is that the excellent fit is accompanied by a lacking mathematical equation to describe the assumed relationship with time. Such an equation is convenient for making predictions. However predictions based on ill-fitting mathematical equations do not mean too much, because of the amount of uncertainty they provide.

References

1. Wavelet (2013) http://en.wikipedia.org/wiki/wavelet. 2 Jan 2013
2. Graps A (1995) An introduction to wavelets. IEEE Comput Sci Eng 2:1–16
3. Merry R (2005) Wavelet theory and applications. A literature study. Eindhoven University of Technology, Eindhoven
4. Addison PS (2005) Wavelet transforms and the ECG: a review. Physiol Meas 26:155–199
5. Fourier J (1822) Theorie analytique de la chaleur. Paris, Edit. by Firmin Didot
6. Anonymous (1997) Haar transform. Stanford Exploration Project. University of Stanford, USA, 11-07

7. Daubechies I (1992) Ten lectures on wavelets. Society for Industrial and Applied Mathematics, ISBN 0-89871-274-2
8. Anonymous (2013). Morlet wavelet. http://en.wikipedia.org/wiki/wavelet. 2 Jan 2013
9. Beylkin G, Coifman R, Rokhlin V (1991) Fast wavelet transforms and numerical algorithms. Comm Pure Appl Math 44:141–183
10. S-plus statistical software (2013) www.s-plus.com. 2 Jan 2013
11. Matlab. Wavelet toolbox. www.matlab.com. 2 Jan 2013
12. ION Script (2004) Wavelets. ION script user's guide. Edit. by Research Systems, Boulder
13. Ishikawa Y (2001) Wavelet analysis for clinical medicine. Edit. by Med Pub, Igaku-Shuppan, Japan
14. Suarez L, Barrett-Connor E (1982) Seasonal variation in fasting plasma glucose levels in man. Diabetology 22:250–253
15. Barrett-Connor E (1980) Factors associated with the distribution of fasting plasma glucose in an adult community. Am J Epidemiol 112:518–523
16. Marmor T, Oberlander J, White J (2009) The Obama administration's options for health care cost control: hope versus reality. Ann Intern Med 150:485–489
17. Cleophas TJ, Zwinderman AH (2012) More on non linear relationships, splines. In: Statistics applied to clinical studies, 5th edn. Springer, Dordrecht, pp 277–288

Chapter 20
Machine Learning and Common Sense

1 Summary

1.1 Background and Objective

This chapter assesses whether receiver operated characteristic (ROC) curve validation of diagnostic tests is accurate.

1.2 Methods and Results

When a diagnostic test uses a validation sample of 100 healthy subjects, and is applied in 1,000 healthy subjects, its accuracy decreases 10 %. This is due to the false positives. Immuno-assays for Lyme disease are 100 % sensitive, but only 70 % specific. The screening of people with fatigue for Lyme has lead to the untenable situation with 10 % of the western population false positive, and has caused serious health damage and deaths in patients due to a mistaken diagnosis.

1.3 Conclusions

ROC validated diagnostic tests tend to lose accuracy when used in real practice.

2 Introduction

Computers are increasingly used in clinical research for making diagnoses and predicting the best possible treatments. However, computers can not think. They can only execute commands as given. One example is the validation of diagnostic

T.J. Cleophas and A.H. Zwinderman, *Machine Learning in Medicine:*
Part Two, DOI 10.1007/978-94-007-6886-4_20,
© Springer Science+Business Media Dordrecht 2013

test. A ROC curve will give you the best cut-off for a true test, either true positive or true negative. However, flexible alphas and betas, otherwise called type I and II errors, otherwise called the chance of a false positive and false negative result, are useful. Without a life threatening illness and with a toxic compound, you may choose an alpha to be small, because you do not wish to treat many healthy subjects with a toxic compound. In contrast, with life threatening disease and no alternative treatment you may prefer a beta to be small, because you do not wish to miss any life threatening diagnoses.

	Disease	No disease
Positive test	a	b
Negative test	c	d

If a, b, c and d are the numbers of patients with respectively true positive, false positive, false negative, and true negative patients, then

$$\text{Sensitivity} = a/(a+c)$$

$$\text{Specificity} = d/(d+b).$$

If $a = 100$, $b = 20$, $c = 0$, and $d = 80$, then the test should have a sensitivity of $100/(100+0) = 1.0$ and a specificity of $80/(80+20) = 0.8$. The proportion of true diagnoses would equal $(a+d)/(a+b+c+d) = (100+80)/(100+20+0+80) = 180/200 = 0.9$. The latter term is often called the overall validity or accuracy of the diagnostic test. A test with 90 % accuracy is excellent, and can be implemented in everyday practice.

In this chapter we will demonstrate that, in everyday practice, a test's accuracy like that of the above test will be far less than expected, and we will use an example from general practice to underscore the clinical relevance of this phenomenon.

3   The Performance of a Validated Test in Everyday Practice

In everyday practice a validated diagnostic test is generally widely used, and many healthy subjects with scanty symptoms are tested. Let us assume that a hospital is going to use the above validated test for assessing 1,000 healthy subjects and 100 diseased patients. From the healthy subjects 80 % (800 subjects) is going to have a negative test, 20 % (200 subjects) a positive one, because the numbers of patients do not usually change the sensitivity and specificity of a test.

If $a = 100$, $b = 200$, $c = 0$, and $d = 800$, then the test should have a sensitivity of $100/(100+0) = 1.0$ and a specificity of $800/(800+200) = 0.8$. The proportion of true diagnoses would equal $(a+d)/(a+b+c+d) = (100+800)/(100+200+0+800) = 900/1100 = 0.8182$. The diagnostic test, although 90 % accurate in the validation procedure, is only 81.82 % accurate in practice. In order to statistically test whether 81.82 % is different from 90 %, a measure of spread is required, and the standard error (SE) of the accuracy can be applied for the purpose. It requires some efforts, but it can be done. Using Bayes' rule and the delta method the following equation is constructed.

$$\text{Var}_{\text{overall validity}} = \text{prev}^2 \times \text{Var}_{\text{sens}} + (1 - \text{prev})^2 \times \text{Var}_{1-\text{spec}} + (\text{sens} - \text{spec})^2 \times \text{Var}_{\text{prev}}$$

$\text{Var} = \text{variance} = (\text{SE})^2$
$\text{prev} = \text{prevalence} = (a+b)/(a+b+c+d)$
$\text{Var}_{\text{sens}} = ac/(a+c)^3$
$\text{Var}_{1\text{-spec}} = db/(d+b)^3$
$\text{sens} = \text{sensitivity}, \text{spec} = \text{specificity}$
$\text{Var} = (a+b)(c+d)/(a+b+c+d)^3$

Using the above equations, the variances of the accuracies are calculated to be 0.0004 and 0.00005, and, in order to test whether 90 % is significantly different from 81.82 %, a t-test (or z-test) is adequate.

$$t = (0.90 - 0.8182)/\sqrt{(0.0004 + 0.00005)} = 0.0882/0.0212 = 4.16,$$

$$p < 0.001.$$

The reason for this difference in accuracy between the laboratory test and the real life test is obvious. When the samples of healthy subjects is ten times larger than the sample of diseased, the number of false positives will increase by a factor 10.

4 Real Data Example

Lyme disease is a tick-borne disease caused by Borrelia infection [1, 2]. A tick of only 2 mm in diameter feeds itself after a painless bite with human blood, and grows during 24 h to 10 mm or more. Then, it is, usually, discovered, and removed. After infection the majority of the patients will develop erythema migrans, consistent of red circles with a diameter up to 75 cm. The infection is, usually, associated with mild inflammatory symptoms. If the diagnosis is suspected, a problem is that cultures are commonly negative, but, fortunately, diagnostic testing based on an immuno-assay for borrelia antibodies is helpful for confirmation of the diagnosis. This is important, because untreated Lyme disease causes serious long-term consequences like neuro-vegetative disturbances and even pareses. The sensitivity of the immuno-assay is 100 %, but the specificity is only 70 % or so [1, 2]. This means, that patients with a negative test are, virtually, excluded from having a Lyme infection. However, about 30 % of the patients without the infection are false positive. If the only consequence would be, a needless antibiotic treatment for these false positives, this is not too much of a problem. However, Lyme testing has been applied in patients with scanty symptoms [3]. The numbers of false positives is currently estimated to be as large as 10 % of the entire western population [4]. These patients are false positive due to previous viral infections like mononucleosis, immunological reasons, or they are just healthy, although with some fatigue [4]. However, they do not suffer from Lyme disease. Patients with a positive test have great difficulty to accept they are healthy, and have been united in patient protection groups worldwide [5]. As commonly observed, patients tend to hear what they *want* to hear, that is, not that they are false positive, but, rather, that the test is positive, although with low accuracy, and this has lead to the situation where many patients with chronic fatigue and a negative test have started to believe they have chronic Lyme too [5–7]. This is supported by widespread misinformation on the internet [1]. This situation where Lyme is used as a diagnosis for patients with chronic fatigue is starting to be untenable, and causes serious health damage and even deaths in patients [1]. All of this might have been prevented, if the immuno-assay had been used for the purpose it was invented for, namely the confirmation of a diagnosis in a patient with the true symptoms of Lyme disease, rather than symptoms of fatigue only.

5   Discussion

The STARD (Standards of Reporting Diagnostic Accuracy) group has launched the quality criteria for diagnostic tests in 2003 [8]. Intervention studies are generally well paid, published in high/impact journals, and provide an excellent career perspective. In contrast, validation studies of diagnostic tests are, generally, not well paid, difficult to publish, provide a poor career perspective, and "post or propter" they are often performed in sloppy ways [9]. Yet, intervention studies are impossible without diagnostic tests, and diagnostic tests are the only real basis of evidence/based medicine. Computers are increasingly used for validation studies of diagnostic tests. However, computers cannot think. They can only execute the commands as given. Also, medical statistics is no bloodless algebra, it requires a lot of biological thinking and a bit of mathematics. It is a discipline at the interface of biology and mathematics. Mathematics is used to answer biological questions. One mathematical principle is the requirement of representative samples to prove your hypothesis is true. The small sample of healthy subjects in a validation study of a diagnostic test is, mathematically, too small to be representative. Well-known principles, that are more biological than mathematical in nature, are the following.

1. The first datum in a situation of ignorance produces the greatest information (e.g., the first case of a disease).
2. Flexible alphas and betas, otherwise called type I and II errors, otherwise called the risks of false positive and false negative patients, is useful. Without a life threatening illness and with a toxic compound choosing a small beta is appropriate, with a life threatening or invalidating condition and no alternative a small alpha is more appropriate.

3. It is recommended to include a safety factor of, e.g., 10 % in your sample size, because it is clinically realistic to account loss of patients while on study due to non-compliance.

We recommend that as a fourth "biological factor" be accounted the loss of accuracy of diagnostic tests, when implemented in everyday clinical practice.
We conclude

1. ROC-curve validation of diagnostic tests overestimates accuracy in real practice.
2. This phenomenon may cause serious health damage and even deaths in patients.
3. Machine learning must be guided by common sense.

6 Conclusions

This chapter assesses whether ROC-curve validation of diagnostic tests is accurate.

When a diagnostic test uses a validation sample of 100 healthy subjects, and is applied in 1,000 healthy subjects, its accuracy decreases 10 %. This is due to the false positives. Immuno-assays for Lyme disease are 100 % sensitive, but only 70 % specific. The screening of people with fatigue for Lyme has lead to the untenable situation with 10 % of the western population false positive, and has caused serious health damage and deaths in patients due to a mistaken diagnosis.

ROC validated diagnostic tests tend to lose accuracy when used in real practice.

References

1. Stanek G, Wormser GP, Gray J, Starle F (2012) Lyme borreliosis, seminar. Lancet 379:461–473
2. Halperin JJ (2011) Lyme disease and evidence-based approach, CAB international book series edit by Halperin, Wallingford
3. British Infection Association (2011) The epidemiology, prevention, investigation and treatment of Lyme borreliosis in United Kingdom patients: a position statement by the British Infection Association. J Inf 62:329–338
4. Coumou J, Van der Poll T, Speelman P, Hovius J (2011) Tired of Lyme borreliosis. Neth J Med 69:101–111
5. Auwaerter PG, Bakken JS, Dattwyler RJ, Dumler JS, Halperin JJ, McSweegan E, Nadelman RB, OÇonnell S, Shapiro ED, Sood SK, Steere AC, Weinstein A, Wormser GP (2011) Antiscience and ethical concerns associated with advocacy of Lyme disease. Lancet Infect Dis 11:713–719
6. Kullberg B, Berende A, Van der Meer J (2011) The challenge of Lyme disease: tired of the Lyme wars. Neth J Med 69:98–100
7. Klempner M, Halperin J, Baker P, Shapiro E, O'Connell S, Fingerle V, Wormser G (2012) Lyme borreliosis: the challenge of accuracy. Neth J Med 70:3–5
8. Bossuyt P, Reitsma J, Bruns D, Gatsonis C, Glasziou P, Irwig L, Moher D, Rennie D, De Vet H, Lijmer J (2003) The STARD statement for reporting studies of diagnostic accuracy, explanation and elaboration. Clin Chem 49:17–37
9. Cleophas TJ, Zwinderman AH (2012) Statistics is no bloodless algebra. In: Statistics applied to clinical studies, 5th edn. Springer, Dordrecht, pp 713–719

Statistical Tables

T-Table

T-Table: v degrees of freedom for t-variable, Q area under the curve right from the corresponding t-value, $2Q$ tests both right and left end of the total area under the curve

$Q = 0.4$	0.25	0.1	0.05	0.025	0.01	0.005	0.001
v $2Q = 0.8$	0.5	0.2	0.1	0.05	0.02	0.01	0.002
1 0.325	1.000	3.078	6.314	12.706	31.821	63.657	318.31
2 .289	0.816	1.886	2.920	4.303	6.965	9.925	22.326
3 .277	.765	1.638	2.353	3.182	4.547	5.841	10.213
4 .171	.741	1.533	2.132	2.776	3.747	4.604	7.173
5 0.267	0.727	1.476	2.015	2.571	3.365	4.032	5.893
6 .265	.718	1.440	1.943	2.447	3.143	3.707	5.208
7 .263	.711	1.415	1.895	2.365	2.998	3.499	4.785
8 .262	.706	1.397	1.860	2.306	2.896	3.355	4.501
9 .261	.703	1.383	1.833	2.262	2.821	3.250	4.297
10 0.261	0.700	1.372	1.812	2.228	2.764	3.169	4.144
11 .269	.697	1.363	1.796	2.201	2.718	3.106	4.025
12 .269	.695	1.356	1.782	2.179	2.681	3.055	3.930
13 .259	.694	1.350	1.771	2.160	2.650	3.012	3.852
14 .258	.692	1.345	1.761	2.145	2.624	2.977	3.787
15 0.258	0.691	1.341	1.753	2.131	2.602	2.947	3.733
16 .258	.690	1.337	1.746	2.120	2.583	2.921	3.686
17 .257	.689	1.333	1.740	2.110	2.567	2.898	3.646
18 .257	.688	1.330	1.734	2.101	2.552	2.878	3.610
19 .257	.688	1.328	1.729	2.093	2.539	2.861	3.579
20 0.257	0.687	1.325	1.725	2.086	2.528	2.845	3.552
21 .257	.686	1.323	1.721	2.080	2.518	2.831	3.527
22 .256	.686	1.321	1.717	2.074	2.508	2.819	3.505

(continued)

T.J. Cleophas and A.H. Zwinderman, *Machine Learning in Medicine: Part Two*, DOI 10.1007/978-94-007-6886-4,
© Springer Science+Business Media Dordrecht 2013

(continued)

	$Q=0.4$	0.25	0.1	0.05	0.025	0.01	0.005	0.001
v	$2Q=0.8$	0.5	0.2	0.1	0.05	0.02	0.01	0.002
23	.256	.685	1.319	1.714	2.069	2.600	2.807	3.485
24	.256	.685	1.318	1.711	2.064	2.492	2.797	3.467
25	0.256	0.684	1.316	1.708	2.060	2.485	2.787	3.450
26	.256	.654	1.315	1.706	2.056	2.479	2.779	3.435
27	.256	.684	1,314	1.701	2.052	2.473	2.771	3.421
28	.256	.683	1,313	1.701	2.048	2.467	2.763	3.408
29	.256	.683	1.311	1.699	2.045	2.462	2.756	3.396
30	0.256	0.683	1.310	1.697	2.042	2.457	2.750	3.385
40	.255	.681	1.303	1.684	2.021	2.423	2.704	3.307
60	.254	.679	1.296	1.671	2.000	2.390	2.660	3.232
120	.254	.677	1.289	1.658	1.950	2.358	2.617	3.160
∞	.253	.674	1.282	1.645	1.960	2.326	2.576	3.090

Chi-Square Distribution

	Two-tailed P-value			
df	0.10	0.05	0.01	0.001
1	2.706	3.841	6.635	10.827
2	4.605	5.991	9.210	13.815
3	6.251	7.815	11.345	16.266
4	7.779	9.488	13.277	18.466
5	9.236	11.070	15.086	20.515
6	10.645	12.592	16.812	22.457
7	12.017	14.067	18.475	24.321
8	13.362	15.507	20.090	26.124
9	14.684	16.919	21.666	27.877
10	15.987	18.307	23.209	29.588
11	17.275	19.675	24.725	31.264
12	18.549	21.026	26.217	32.909
13	19.812	22.362	27.688	34.527
14	21.064	23.685	29.141	36.124
15	22.307	24.996	30.578	37.698
16	23.542	26.296	32.000	39.252
17	24.769	27.587	33.409	40.791
18	25.989	28.869	34.805	42.312
19	27.204	30.144	36.191	43.819
20	28.412	31.410	37.566	45.314
21	29.615	32.671	38.932	46.796

(continued)

(continued)

df	Two-tailed P-value			
	0.10	0.05	0.01	0.001
22	30.813	33.924	40.289	48.268
23	32.007	35.172	41.638	49.728
24	33.196	36.415	42.980	51.179
25	34.382	37.652	44.314	52.619
26	35.563	38.885	45.642	54.051
27	36.741	40.113	46.963	55.475
28	37.916	41.337	48.278	56.892
29	39.087	42.557	49.588	58.301
30	40.256	43.773	50.892	59.702
40	51.805	55.758	63.691	73.403
50	63.167	67.505	76.154	86.660
60	74.397	79.082	88.379	99.608
70	85.527	90.531	100.43	112.32
80	96.578	101.88	112.33	124.84
90	107.57	113.15	124.12	137.21
100	118.50	124.34	135.81	149.45

F-Distribution

df of denominator	2-tailed P-value	1-tailed P-value	Degrees of freedom (df) of the numerator												
			1	2	3	4	5	6	7	8	9	10	15	25	500
1	0.05	0.025	647.8	799.5	864.2	899.6	921.8	937.1	948.2	956.6	963.3	968.6	984.9	998.1	1017.0
1	0.10	0.05	161.4	199.5	215.7	224.6	230.2	234.0	236.8	238.9	240.5	241.9	245.9	249.3	254.1
2	0.05	0.025	38.51	39.00	39.17	39.25	39.30	39.33	39.36	39.37	39.39	39.40	39.43	39.46	39.50
2	0.10	0.05	18.51	19.00	19.16	19.25	19.30	19.33	19.35	19.37	19.38	19.40	19.43	19.46	19.49
3	0.05	0.025	17.44	16.04	15.44	15.10	14.88	14.73	14.62	14.54	14.47	14.42	14.25	14.12	13.91
3	0.10	0.05	10.13	9.55	9.28	9.12	9.01	8.94	8.89	8.85	8.81	8.79	8.70	8.63	8.53
4	0.05	0.025	12.22	10.65	9.98	9.60	9.36	9.20	9.07	8.98	8.90	8.84	8.66	8.50	8.27
4	0.10	0.05	7.71	6.94	6.59	6.39	6.26	6.16	6.09	6.04	6.00	5.96	5.86	5.77	5.64
5	0.05	0.025	10.01	8.43	7.76	7.39	7.15	6.98	6.85	6.76	6.68	6.62	6.43	6.27	6.03
5	0.10	0.05	6.61	5.79	5.41	5.19	5.05	4.95	4.88	4.82	4.77	4.74	4.62	4.52	4.37
6	0.05	0.025	8.81	7.26	6.60	6.23	5.99	5.82	5.70	5.60	5.52	5.46	5.27	5.11	4.86
6	0.10	0.05	5.99	5.14	4.76	4.53	4.39	4.28	4.21	4.15	4.10	4.06	3.94	3.83	3.68
7	0.05	0.025	8.07	6.54	5.89	5.52	5.29	5.12	4.99	4.90	4.82	4.76	4.57	4.40	4.16
7	0.10	0.05	5.59	4.74	4.35	4.12	3.97	3.87	3.79	3.73	3.68	3.64	3.51	3.40	3.24
8	0.05	0.025	7.57	6.06	5.42	5.05	4.82	4.65	4.53	4.43	4.36	4.30	4.10	3.94	3.68
8	0.10	0.05	5.32	4.46	4.07	3.84	3.69	3.58	3.50	3.44	3.39	3.35	3.22	3.11	2.94
9	0.05	0.025	7.21	5.71	5.08	4.72	4.48	4.32	4.20	4.10	4.03	3.96	3.77	3.60	3.35
9	0.10	0.05	5.12	4.26	3.86	3.63	3.48	3.37	3.29	3.23	3.18	3.14	3.01	2.89	2.72
10	0.05	0.025	6.94	5.46	4.83	4.47	4.24	4.07	3.95	6.85	3.78	3.72	3.52	3.35	3.09

10	0.10	0.05	4.96	4.10	3.71	3.48	3.33	3.22	3.14	3.07	3.02	2.98	2.85	2.73	2.55
15	**0.05**	**0.025**	**6.20**	**4.77**	**4.15**	**3.80**	**3.58**	**3.41**	**3.29**	**3.20**	**3.12**	**3.06**	**2.86**	**2.69**	**2.41**
15	0.10	0.05	4.54	3.68	3.29	3.06	2.90	2.79	2.71	2.64	2.59	2.54	2.40	2.28	2.08
20	**0.05**	**0.025**	**5.87**	**4.46**	**3.86**	**3.51**	**3.29**	**3.13**	**3.01**	**2.91**	**2.84**	**2.77**	**2.57**	**2.40**	**2.10**
20	0.10	0.05	4.35	3.49	3.10	2.87	2.71	2.60	2.51	2.45	2.39	2.35	2.20	2.07	1.86
30	**0.05**	**0.025**	**5.57**	**4.18**	**3.59**	**3.25**	**3.03**	**2.87**	**2.75**	**2.65**	**2.57**	**2.51**	**2.31**	**2.12**	**1.81**
30	0.10	0.05	4.17	3.32	2.92	2.69	2.53	2.42	2.33	2.27	2.21	2.16	2.01	1.88	1.64
50	**0.05**	**0.025**	**5.34**	**3.97**	**3.39**	**3.05**	**2.83**	**2.67**	**2.55**	**2.46**	**2.38**	**2.32**	**2.11**	**1.92**	**1.57**
50	0.10	0.05	4.03	3.18	2.79	2.56	2.40	2.29	2.20	2.13	2.07	2.03	1.87	1.73	1.46
100	**0.05**	**0.025**	**5.18**	**3.83**	**3.25**	**2.92**	**2.70**	**2.54**	**2.42**	**2.32**	**2.24**	**2.18**	**1.97**	**1.77**	**1.38**
100	0.10	0.05	3.94	3.09	2.70	2.46	2.31	2.19	2.10	2.03	1.97	1.93	1.77	1.62	1.31
1000	**0.05**	**0.025**	**5.04**	**3.70**	**3.13**	**2.80**	**2.58**	**2.42**	**2.30**	**2.20**	**2.13**	**2.06**	**1.85**	**1.64**	**1.16**
1000	0.10	0.05	3.85	3.00	2.61	2.38	2.22	2.11	2.02	1.95	1.89	1.84	1.68	1.52	1.13

Index

Printed in the United States
By Bookmasters